Kishor Bhagwati

Managing Safety

Managing Safety. Kishor Bhagwati

ISBN 3-527-31583-7

Further of Interest

Schütz, H., Wiedemann, P. M., Hennings, W., Mertens, J., Clauberg, M.

Comparative Risk Assessment

2006
Hardcover
ISBN 3-527-31667-1

Budde, F., Felcht, U.-H., Frankemöller, H. (Eds.)

Value Creation

2nd, Completely Revised and Extended Edition
2006
Hardcover
ISBN 3-527-31266-8

Bamfield, P.

Research and Development in the Chemical and Pharmaceutical Industry

3rd, Completely Revised and Extended Edition
2007
Hardcover
ISBN 3-527-31775-9

Kishor Bhagwati

Managing Safety

A Guide for Executives

WILEY-VCH Verlag GmbH & Co. KGaA

The Author

Kishor Bhagwati
International Management Consultant
25 Chemin de Plantaz
1095 Lutry
Switzerland
bhagwati@bluewin.ch.

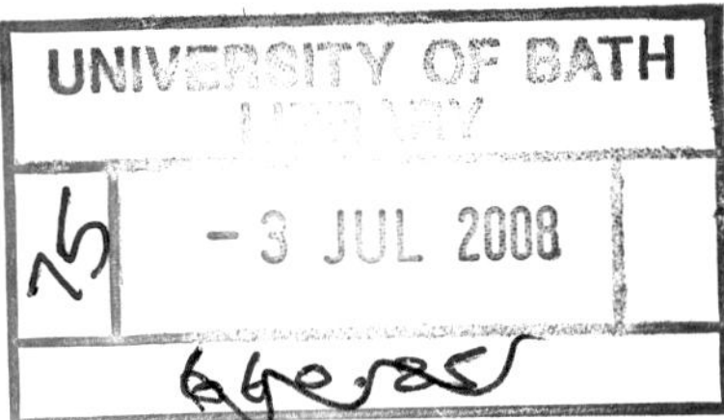

Library of Congress Card No.: applied for

British Library Cataloging-in-Publication Data:
A catalogue record for this book is available from the British Library.

Bibliographic information published by the Deutsche Nationalbibliothek
The Deutsche Nationalbibliothek lists this publication in the Deutsche Nationalbibliografie; detailed bibliographic data is available in the Internet at http://dnb.d-nb.de.

Typesetting TypoDesign Hecker GmbH, Leimen
Printing betz-druck GmbH, Darmstadt
Binding Litges & Dopf GmbH, Heppenheim
Cover Design A-Plus-Design, Achim Bauer, Ludwigshafen

Printed in the Federal Republic of Germany
Printed on acid-free paper

ISBN-13: 978-3-527-31583-3
ISBN-10: 3-527-31583-7

Table of Contents

Managing Safety. Kishor Bhagwati

ISBN 3-527-31583-7

Preface

Safety was never my primary interest during my studies as a chemical engineer (although I got sprayed once with concentrated sulfuric acid during an experiment, making my face look like a moonscape), nor when I started working in the industry. I found myself very soon in the normal rut of thinking that safety was the job of the safety professional, and had nothing to do with us production people. As an engineer I was trained to see that the machines and apparatus one designed ran without a hitch, that they complied with the codes of practice so that they would not burst or explode, and that a plant layout was designed to optimise production performance. The safety of the people working in these plants was of secondary importance, and was best left to the safety people.

One day, one of the workers working in a plant for manufacturing parathion, a highly poisonous insecticide, of which I was the plant manager, suffered a bad accident. He was reading a rotameter for measuring the flow of the pesticide. A rotameter is a slightly conical glass tube marked with a scale and mounted vertically with a metal float inside that rises up or down depending on the amount of flow. A worker, who was in the habit of opening his mouth when trying to read something closely, was noting down the position of the float, when suddenly the glass tube burst and sprayed the poison into his mouth.

After first-aid by the company doctor he was immediately transferred to the hospital. He was in a very critical condition, suspended between life and death, and was at last out of danger after six hours of treatment. I waited at his bedside all the time, seeing his cramps and his fight with death. And I started thinking. I thought of the man in bed who had come to work to provide for his family, and what his death would mean to them. I thought of the trust he had put in me, his manager, doing unquestioningly the work I had told him to do and

Managing Safety. Kishor Bhagwati

ISBN 3-527-31583-7

believing that he would not get hurt working for me, because I would have taken care to see that he got back to his family in the evening as healthy as he had come to my plant. This was the time when safety shifted in my mind from being a purely technical problem for specialists, to something emotional, something in which I had to involve myself to protect the health of those who came to me. Safety became humanised for me.

I was fortunate to work in companies that had very high standards of safety, and I went through my career believing that this was the way most companies worked. When I started consulting in safety management and visited companies of repute, I was astonished that this was not the case, and that the managers in these companies still stuck to the old idea of safety being someone else's job. I am not talking of small companies, but of really big ones with international operations. Surprises awaited me at every corner. Many of these I have used as examples in this book. I often had to control and conceal my irritation when I talked to managers in these companies about safety, because the answers they gave were so disrespectful of workers. I had thought those times had passed a century ago, but no, they were and are still living in the minds of many a managers.

At all my consulting projects, where I could get top managers to involve themselves in the safety effort, I was greatly satisfied to see the change in their attitude after some time, and the accident rates that started dropping. I spread my message with an intensity not usually employed in consulting, because I was convinced that what I had learned that day at the bedside of my worker was so fundamental in leading people, that I wanted all others who are leaders to know about it and to act accordingly.

This book talks only about occupational safety, but the principles evolved here are applicable to any type of management. I have been a manager and have had to deal with all types of managers during my career in the industry and later as consultant. The truths gleaned in this book are universal. At the bottom of them is a new relationship between the workers and their managers built on mutual respect and – should I say it? – liking. I always respected and liked my workers and I was never ashamed of showing it to them. And they have returned these sentiments to me with compound interest.

One day my wife, with whom I used to discuss all my ideas and problems, suggested that I write down all the things that I tell others.

That is how this book was born. It was the first time I was writing something that was not an annual report, a project report, an audit report or a consulting report. I was writing down my thoughts and my beliefs. I often thought: who would be interested in reading them? But my wife Heidi and my children Sandeep, Anjana and Shushila showed confidence every time I had such thoughts, and gave me the courage to complete it. To them this book is dedicated.

At Wiley-VCH, my special thanks go to Dr Stefan Pauly, who accepted my book proposal, Mrs Karin Sora, whom I presented my concept orally, Dr Rainer Münz, who took over the project of publishing the book, and Mr Peter Biel, who was in charge of production.

Lutry / Lausanne, Switzerland *Kishor Bhagwati*
March 2006

1
Introduction

Dear Fellow Manager

Having been there where you are today, I know well how much you have invested to get there. You studied, slogged nights to pass your exams, entered the world of work nervously, and had a lot of fun and satisfaction doing the things you did.

You learned to live with all those big and small snags of politics and to recognise the hurdles on your way to a better position and more money. You learned not to stumble over them, but to overcome them to reach your coveted goal of becoming a manager.

You became an expert in discovering problems and dealing with them, your intelligence and quick grasp of the situation helping you climb always higher and higher, from stepping-stone to stepping-stone of junior manager, senior manager, area manager, and so forth, till some of you reached the position of a works or site manager, or even higher.

Did it require a lot of effort and sacrifice? It sure did! Your families craved for more time with you, but you were *so* busy. You even brought work home to catch up with the backlog produced by all those lengthy (and often boring and ineffective) meetings at the office. You had to travel extensively, getting irritated about plane delays and uncomfortable airport lounges, eating too much at business dinners, and coming home dead tired to do anything else but recover from these chores, with the Damocles sword of work collected at the office during your absence hanging over your thoughts.

Now here you are, at one of the top rungs, managing the problems of production, raw materials, quality, and maybe also sales and marketing, but above all, profits. Profits for the company that has entrusted you with running the show, however big or small it may be. You have to worry about production stoppages, quality noncompliance,

Managing Safety. Kishor Bhagwati

ISBN 3-527-31583-7

worker problems, union matters, you name it. One would often like to be like one of those Indian deities with ten or twenty arms and enough time to do all that can be done with them.

Is all the work listed, or is something missing? Where was the safety of your people in the list?

But of course, you care about safety. Haven't you employed safety professionals to do this job? Good, well-trained professionals who should look everywhere with their trained eyes, carry out safety audits, improve working conditions, catch people not doing things safely, investigate accidents to find the culprits, present you regularly with safety statistics (which you hardly have time to study) that hopefully prove that safety is improving and accident figures are going down? As a matter-of-fact, you have even seen to it that the safety professional reports directly to the HR manager or sometimes even to you! You approve sufficient money (not too much, of course) for his plans and projects to improve the safety of machines! That is how important you consider safety is. You are overloaded with so much other work that you cannot have time to worry about safety personally, and are happy that there is someone who takes the load off your shoulders. Can one take safety more seriously?

Yes, indeed!

As a manager, one is a juggler. You juggle various balls like productivity, quality, raw materials, personnel, customer satisfaction, etc. You cannot afford to let any of them fall down. Once in a while one of the balls reaches the apex and gets your highest attention. But even then you continue to juggle other balls, not missing any.

Now you need to add one more ball to those you juggle. And that ball is: the safety of your workers.

You are great at fulfilling your obligation towards the company by running a factory smoothly. You are where you are because you have proved that, in spite of ups and downs, you have managed to keep the production running and the profits flowing in. Those accidents that happen are just a bother, causing unnecessary trouble. Just because some idiot on the shop-floor does not pay enough attention, the production has to be stopped. What the hell is your safety guy doing? Why can't he stop all these silly accidents? High time you started looking into his performance rating.

You better start looking into a mirror. The safety of your workers is as much your responsibility as all other responsibilities you bear. The worker trusts you to take care of him. He comes to work expecting to go home uninjured. You are responsible for providing him a place of work where he knows he is safe and will not suffer an injury. His safety, believe it or not, is your job, and not that of the safety professional.

Let us come back to your juggling. Worker safety is one of the balls you have to juggle with the same dexterity as all other balls. But there is a difference. The safety ball is made of glass. If any of the other balls falls, you can work at bringing it back into position. Not so with the safety ball: if it falls, it breaks, and there is no way to repair it, neither with all the King's horses, nor with all the King's men. There is nothing you can do to give back to a worker what he has lost, neither his eye, nor his arm, and never his life.

Just as you would never fully delegate other important managerial aspects to your subordinates, so also you can not fully delegate the safety of your people to somebody else. Can you imagine leaving the control over all financial aspects to your financial manager, looking at it once in a while in a cursory manner? Would you not constantly like to know what is happening in real time and immediately act when the profits show a downward trend? Would you not have sleepless nights when something affecting profits gets out of control? Yes, you would, because it is the responsibility you bear.

And you bear the same responsibility for the safety of your workers. You often proclaim that your workers are your "most valuable assets«. If you really believe that, then start putting worker safety where it belongs, viz. on an equal level with all your other managerial responsibilities. There are many reasons for doing this, apart from the caretaker function you don. Fewer accidents mean less interruption in the plant, resulting in higher productivity, better worker morale and higher quality. These will be clearly reflected in your bottom line.

I think we now agree that safety needs greater attention than it is getting today. You, as the mover of a company's performance, have to involve yourself in the act. The question is: how? Nobody taught you in all your studies and career how to fulfil this obligation in an optimum manner, slicing away a minimum of time from your other responsibilities that already have given you 120% work. But it can be done. And my experience has shown that the satisfaction in the success in fighting injuries is overwhelming and highly satisfying.

At a refinery in Western Europe, which had on the average about 110 accidents per year, introduction of correct safety-management techniques brought accidents tumbling down in a few years. At the beginning of January last year, five years after the safety project was started, the works manager called me and said, "I have been at this site for 23 years, and works manager since 9 years. In all my 23 years there never was a finer moment than on the 2nd of January this year, when my safety chief came to me and told me: 'Sir, we had zero accidents last year!'"

This book will show you how you too can get this wonderful feeling of having reached a goal that is worth striving for. Chapter by chapter you shall come to a deeper understanding of what causes accidents, why they happen, and how you, as a manager, can prevent them. You will learn how, with a minimum investment of time and hardly any money, you too can achieve supreme safety. There may be many ways of reducing costs and gaining higher productivity, but none is finer and more humane than reducing accidents and injecting your people with a strong motivation that comes from the feeling of being in caring hands.

Let us start on this journey to an injury-free workplace. *Powered by You!*

And remember:

Take care of your people, and they will take care of your profits.

2
Consequences of Accidents

Before you learn how to reduce accidents in your area of management, you have to give some thought to what an accident really means, and what effects and consequences it has on different groups of people and on the company. The consequences of an accident are widespread and affect areas one does not readily expect at first sight. You may have heard that accidents cost money, but how much and to what extent, may not be known to you. Once you realise what can result from an accident, you will start thinking differently about them and see a definite purpose in reducing them as far as possible. In this chapter we want to look deeper into these consequences, and why they need to be controlled.

Who and what is affected by an accident?

1. The victim himself
2. His family
3. His colleagues
4. His superiors
5. The worker morale, and
6. The company business.

Let us go through these six points one by one.

1. The victim himself

The immediate effect is the bodily pain. This can be slight or very strong. Then come the subsequent pains and suffering from the injury. If the accident has resulted in a loss of a body part, then, besides the bodily pain, there is the maiming that lasts a whole lifetime. His quality of life reduces markedly, destroying the dreams and plans he

Managing Safety. Kishor Bhagwati

ISBN 3-527-31583-7

had for his future life. If the injury has resulted in him being bound to a wheel-chair, his mobility is restricted, and if he becomes blind, then he becomes totally dependant on others to nurture him. Often, in such cases, the family breaks apart.

Even smaller accidents could result in such grievous situations. For example, a person may slip so badly, that he breaks his back and is lame from the hip downwards, or falls from a ladder and breaks his neck. Because no one can guarantee how big or small an injury might be, we have to assume the worst possible consequences in every accident.

2. His family

For his immediate family it is a great shock. They suffer for the victim, feel his pain, have to take care of him after his leaving the hospital, have to readjust their lives, both physically and financially, and may face ruin if the disablement is permanent.

If you would really like to convince yourself and your people of the consequences associated with an accident, get your training department to purchase the video "Remember Charlie" and watch it together with your staff in your management meeting[1)]. Charlie Morecraft was an employee of Exxon, who was critically injured in an accident that resulted in burns to over 45% of his body. Charlie just tells his story in this film, and when he tells it, the audience sits in rapt attention as they are moved to a new level of awareness. Charlie has that rare ability of creating vivid images with ordinary words. His story is unforgettable. He tells us things we already know, but he tells them in a way that will dramatically change our attitudes toward safety for ever.

His primary message is: "Safety is about going home at the end of the day, kissing your wife and hugging your kids". Watch this video, and make all the managers at your site watch it. Before one starts working on improving safety, one has to know its emotional background, and this video will give it to you.

1) Suppliers under Google searchword: "Remember Charlie"

3. His colleagues

Have you ever been in a plant that had a fatality? You will find the colleagues of the victim at a loss and unable to concentrate on work, and often psychiatric help is required for them. It is very disturbing for the victim's colleagues to digest the fact that someone they knew close enough, and was their friend, is there no more. For them it is almost like losing a close relative. They start asking themselves: Is the machine I have to work on really safe? Could I be the next victim? Insecurity, sadness and demoralisation prevail.

It need not be a fatality; serious accidents result in the same type of demoralisation. Even in the case of minor accidents, if they are frequent, workers start blaming the management for not taking care of their safety better. Grumbling starts, and with that loss of faith and deterioration of worker-manager relationship. During my conversations with workers in factories with a noticeable accident rate, I always heard "What are we to the management? We are just wheels in a machinery here to ensure their production. Who cares for us? We are nothing to them." And then a barrage of complaints of how heartless the managers are starts.

4. His superiors

He is the one in line management whom it hits first. He has to arrange all the activities resulting from the accident, e.g. seeking replacement for the injured worker, arranging for his transport to a medical facility, carrying out the accident investigation, writing accident reports to authorities, and, in case of a serious accident, answering the police and the public prosecutor. Apart from this, he has to calm the colleagues of the worker and reinstil in them the work spirit so that the production goes on. I have not met a single manager who has not felt sorry for his man who has been hurt, and who hasn't asked himself, why did it have to happen, and why to him? He may be convinced that it was the worker's fault, but still the questions do not leave his mind and the sadness remains.

5. The worker morale

Good worker morale is the most important driving force you will need as a manager if you want to achieve superior performance in any field, not only in safety. Only satisfaction at the workplace can produce in the workers the wish to do more than is legally required of them, and to give their best. No amount of money can ever achieve this. And, as mentioned before, frequent accidents, however and by whomever they are caused, give workers the feeling that they are not being taken proper care of. People may not love their boss, but they expect him to be fair, just and *caring*. If this faith is missing, one automatically gets sloppy workmanship, poorer quality and bitterness in the workforce. Every accident, therefore, has a substantial impact on the worker morale.

6. The company

The image a company has in the eyes of its neighbours and customers is very important for business. In the eye of its customers, a company with a high accident rate is a badly managed company. And that is what it really is! Due to frequent accidents the company suffers delivery problems, quality problems and cost (productivity) problems. Any customer who intends to make you his preferred or sole supplier will not tolerate frequent delivery interruptions. On the other hand, a company with excellent safety record generates an impression of managerial excellence in the minds of its customers. A parallel can be drawn to the safety of airlines. With which airline would you prefer to fly? The one that has a reputation of being a well-managed airline with excellent ground service and nearly no accidents, or the one that is poorly managed, has incompetent ground service and has frequent problems, including crashes?

Let us now discuss the costs of accidents. The only costs you may be seeing are the visible, direct costs, which are covered to a large extent by the insurance companies. Depending on local customs, the direct financial consequences of accidents could be either your participating in the costs of the accident, or in an increase in the insurance premium you have to pay, or in a reduction of your rebate for a below-average accident rate. These are the directly visible costs. But what about the indirect costs? Unfortunately, our book-keeping systems do not separate these costs from other costs, and thus we are led to believe that the costs of the accidents are not high enough to justify investment of any kind. All these indirect costs are submerged under various headings and overheads. The fact, however, is: they are there! Let us try listing where some of these hidden indirect costs come from:

- Lost time of injured and his fellow worker(s)
- Replacement worker, his cost and training
- Damaged equipment and its replacement
- Lost service time of equipment
- Lost time of production
- Damage to plant and its repair
- Spoiled product
- Accident investigation time
- Associated administrative time
- Downtime during investigation
- Failure to keep deadlines
- Adverse publicity
- Loss of customer
- Et cetera.

Figure 1 shows an iceberg representing actual costs of accidents. The visible costs, above the water level, are much less than the hidden costs below the water level. These are probably the costs you have been looking at when deciding whether an investment in the field of safety is justified, or not. Hidden from your view are the manifold costs of items listed above, covered unknowingly and innocently under the blanket of departmental overheads. However invisible they may be, the fact is that they affect the bottom line as much as any other factor gone astray in the company. It is essential that one starts looking at them more seriously. Putting your head in the sand will not make them go away; all that we will get is sand in our mouth making us grit our teeth when the realisation comes.

H. W. Heinrich first proposed a relationship between these costs in his book "Industrial Accident Prevention" published in 1931 (last edition 1980 by McGraw Hill). Heinrich carried out an empirical investigation of about 550 000 insurance accident claims and came to the conclusion that for each $1 spent on insurance premiums, the indirect

Figure 1 The Cost Iceberg

costs in an accident were $4. His ratio of visible to invisible costs was thus 1:4. In the 1990s, the UK Health & Safety Executive (HSE) carried out a large-scale study[2] of several industries, and concluded that the ratios were between 1:8 and 1:36, much higher than 1:4 determined by Heinrich 60 years earlier.

But worse is yet to come.

2) "The Costs to the British Economy of Work Accidents and Work-related Ill Health", HMSO 1994.

These costs are not to be deducted from sales, but from profits. Depending on your ratio of sales to profits, each accident that has reduced your profit has eaten up an equivalent amount of your sales. The Occupational Safety and Health Administration (OSHA) of the US Department of Labor once gave an example to understand how this looks in real life:[3)]

To pay for an accident with a direct cost of only USD[4)] 1000:

- A soft drink bottler would have to bottle and sell over 61 000 cans of soda.
- A food packer would have to can and sell over 235 000 cans of food.
- A bakery would have to bake and sell over 235 000 donuts.
- A contractor would have to pour and finish 3000 square feet (300 m^2) of concrete.
- A ready-mix company would have to deliver 20 truckloads of concrete.
- A paving contractor must lay 900 feet (270 m) of a two-lane asphalt road.

One would be extremely fortunate if all the accidents that took place at a site would cost around USD 1000. In fact, the average cost of an accident, taking the whole range of accidents from the minor cuts to the major ones requiring hospitalisation for several months, is much higher.

OSHA has published Cost Calculation Worksheets in its Safety & Health Management Systems eTool[5)]. The first table calculates the annual accident costs based on the types of accidents, and the second table lets one calculate the impact of accidents on profits and sales. In the internet publication, the unit costs in the third column in Table 1 are based on US National Safety Council's 1998 figures. I have obtained the 2004 figures from them[6)] and have used them here. For the sake of illustration, I have put numbers in the "Enter" field for a company with a good safety record.

3) "$afety Pays", published by the U.S. OSHA, 1996.
4) Published in 1982 with a figure of USD 500, here adjusted to the equivalent value of the dollar in 2004.
5) Internet: http://www.osha.gov/SLTC/etools.
6) Private correspondence, November 2005.

Table 1 Cost of accidents

U.S. Department of Labor
Occupational Safety & Health Administration

Safety & Health Management Systems eTool
Cost Calculation Worksheet

I. ESTIMATED ANNUAL ACCIDENT COSTS

	Enter		Results
Annual number of occupational deaths	0	× $1,110,000* = $	0
Annual number of lost workday cases	10	× $38,000* = $	380,000
Annual number of reportable cases without lost work days	30	× $9,000* = $	270,000
Total estimated annual cost of occupational deaths, injuries and illnesses		= $	**650,000**

Calculate

*Includes both direct and indirect costs, excludes property damage.

Table 2 Impact on profits and sales

II. IMPACT OF ACCIDENTS ON PROFITS AND SALES
(Compare accident costs to company profits)

Sales Volume: $ 100,000,000 (Enter)

Profit Margin: 5 % (Enter)

Sales Volume × % Profit = **Annual Profits: $** 5,000,000 (Result)

Accident Costs (Estimated or Actual): $ 650,000 (Result or Enter)

Accident Costs as a Percent of Profits: 13% (Result)

****Amount of Sales Needed to Replace Lost Profits**: $ **13,000,000** (Result)

(If profit margin is 5%, then it takes $20 of sales to replace every dollar of loss)

Calculate

** Not all indirect costs will impact a company's bottom line on a dollar for dollar basis. Therefore, using all indirect costs within the calculation overestimates the amount of sales needed to replace lost profits. Likewise, only using direct costs in this comparison would underestimate the amount of sales needed. Actual costs will be somewhere in between the two figures.

Astounding? Shocked? Did you ever imagine that with a 5% profit margin and only 10 lost workday cases in a company with sales of about $100 million the cost of accidents alone would be $650 000, eating up sales to the tune of over $10–13 million? And this is **not** counting property damage.

These are the facts, whether you like to believe them, or not.

Can you imagine how profitable it would be in the above case if the accidents were reduced, say, by 50%? It would have equalled an additional profit of $325 000, equivalent to additional sales of $6.5 million!

Wanting to check how accident costs look in Europe, I had initiated a study by the University of Lausanne School of Economics in Switzerland to determine the direct and indirect (hidden) costs in my company's factories in three European countries with differing compensation systems, viz. UK, France and Germany[7]. The result was equally flabbergasting. With a total of 360 recordable cases (with and without days away from work), the sum of direct and indirect costs turned out to be $13.5 million, giving the cost per accident of $37 500. Europe, with its extensive social net, turns out to be more "expensive" than the US

Whatever your motive, I am sure you now agree that reduction of injuries is an aim worth striving for. This book will show you how, with minimum time and effort on your part, you can make the area under your management an injury-free workplace.

In the cost-cutting times we are living in, proper safety management shows a new way to reduce costs without taking resort to the commonly used retrenchment of people. A way that is definitely more civilised, caring and humane.

7) Dissertation Christian Vaney, Université de Lausanne, Ecole des Haute Etudes Commerciales.

3
A Small Experiment

The whole science of occupational safety, as discussed in this book, is based on two basic truths. Instead of my writing them down, I would like you to arrive at them yourself. To do this, we shall carry out a small experiment that just needs a piece of paper and something to write. It is not a quiz, but there is an invaluable prize that you will win, a prize of insight and knowledge.

The experiment can also be carried out in a group. I recommend you do it at your management meeting with your managers who report directly to you, and ask that they subsequently carry it out with their subordinates when you start your company's or factory's journey to excellence in safety. It takes only a few minutes, but these are minutes that will bring you the recognition of the most basic concepts of accidents and their causes.

Step 1

Think of an accident you know. It should be an accident with injury. It can be from the working environment, or from the private area. It could be a serious accident at the plant or an accident at home or while driving. You need not be directly involved in it, but you should have first-hand knowledge about it. It has to be an accident of which you know some details, not just something you have read about.

One condition, though, applies. We shall not be considering accidents of children and aged persons. Also not sports accidents. The reason will be clear to you at the end of the experiment.

Write down the facts of the accident you have chosen in about two to three lines. No detailed description is required, only the basic facts. Just describe the situation in which the accident happened and the degree of injury. If you are carrying out the experiment alone, you can

Managing Safety. Kishor Bhagwati

ISBN 3-527-31583-7

write it in the form of Figure 2. If you are doing this experiment in a group, then ask everybody to write it on a piece of paper, or copy the form for them. About three minutes should suffice for this step. Nobody will be asked to read aloud what he wrote.

Step 2

Now try to determine the cause of the accident. Not the immediate cause, but the deeper cause, the so-called root cause. The root cause lies below all the superficial causes, and can be found only by questioning the cause arrived at first thinking. For example, a worker in a chemical plant slipped on a wet floor and fell down, hurting himself enough to stay home for one day. The normal reasoning would be: the worker is not to blame, because he could not know that the floor was wet. Measures to prevent a recurrence therefore: none. It was just one of those unavoidable accidents that happen. C'est la vie!

Digging deeper means asking the question: Why was the floor wet? The answer could be: Because a colleague had just washed the floor after cleaning a spill of a hazardous chemical. The next question then would be: Why did he not put up a sign or a barrier warning others about the wet floor?

Something the superficial investigation showed as not having any cause showed on further questioning to be the unsafe act of a colleague, who forgot or did not think it necessary to put up a sign or a barrier.

Analyse your accident accordingly, and answer just one question: Was the real cause of the accident a human act, or a mechanical failure? In other words, was it an unsafe act of the victim himself or somebody else, or an unsafe condition? Put your cross in the appropriate box on the form.

Step 3

In the last step, just answer the following question:
Could anything have been done – either by the victim or someone else – three minutes, three hours, three days or three months ago that would have prevented this accident? Just answer yes or no, and put your cross in the appropriate box.

Analysis of an Accident	
Details of the accident:	
Root cause: Unsafe act ☐	Root cause: Unsafe condition ☐
Preventable: Yes ☐	Preventable: No ☐

Figure 2 Analysis of an Accident

Results

Let us take the first part of the experiment.

If you have carried out this experiment alone, with all probability you would have put your cross in the box for "unsafe act". If not, go deeper into your analysis, and see whether there is not an act of a human being behind it. For example, someone trips over a hose on the floor. The hose lying there is, of course, an unsafe condition. But did the hose unreel itself and lay itself down in the way? Did it fall from heaven and position itself so that someone would trip over it and fall? No. The answer is: someone used that hose, and forgot to wind it back out of the way. The unsafe condition was caused by the unsafe act of someone else. The root cause, therefore, is an unsafe act.

If you are doing this experiment in a group (say, your management committee), ask them to raise their hand to show how many thought the root cause was an unsafe act. You will find at least 90% raising their hands. If someone does not, ask him the same question as I asked you before. Ask him, how did the unsafe condition arise? Apart from natural catastrophes, it is always due to somebody's unsafe act.

Once when I was giving a public lecture and carrying out this exercise, one of the participants said that he cut his hand due to an unsafe condition. I asked him how it had happened. He said that he tried to remove shards of a glass that had fallen out of his hand into the wash basin and broken. He cut his hand badly doing it. He called the existence of shards in the wash basin an unsafe condition. I asked him whether he could describe what exactly did he do. He said; "Well, I tried to pick up the shards with my hand ...." He suddenly stopped talking in the middle of the sentence, and I did not have to say anything more.

The conclusion of the first part of the experiment is:

Nearly all accidents happen due to human beings carrying out unsafe acts, and very few happen due to unsafe conditions not generated by humans.

However large a group you carry out the experiment with, and however often, you will always end up with the same result. Remind yourself and others that the accident was selected by them, was analysed by them and they arrived at the conclusion themselves. Nobody asked them to analyse a "given" accident, and none was prompted or influenced in his thinking.

Why did we exclude children, aged persons and sportsmen? Because the answer would have been too obvious and simple. Children would do things out of ignorance, aged persons out of lack of coordination, and sportsmen voluntarily. Children do not know what risk is, aged persons cannot recognise risks and cannot control their movements to avoid it, and sportsmen do know the risk, but take it willingly, because, as the saying goes: "No risk, no fun!"

I live in Switzerland, a country famous for its snowy mountains and winter sports (and, of course, the best chocolates, cheeses and those marvellous watches). During the 6 months of the skiing season every year, about 100 000 accidents with injuries that need medical treatment are registered with the insurance companies. The fatality count is about 30 per year[8]. If you had a factory with numbers like that, it would be closed down by the authorities immediately. But people still go on coming to the mountains, standing for hours at the lifts and willingly facing the risk of getting injured. No risk, no fun!

Statistics show that the fatality rate is highest in mountaineering. In Switzerland alone on average 65 fatalities are registered every year.

8) Source: Swiss Council for Accident Prevention – CAP, Bern.

Top of the list of mountaineering accidents are the Himalayas. In the 52 years since Sir Edmund Hillary of New Zealand and the Sherpa Tenzing Norgay of Nepal climbed to the top of Mount Everest in June 1953, about 2560 persons have attempted the climb. Of these, about 200, i.e. nearly 10%, died on the way[9]. This has not stopped mountaineers still attempting a climb. Registrations for climbing permits are coming in daily.

The second part of the experiment must have resulted in 100% of the participants saying: Yes, the accident could have been prevented. All things that happen are consequences of things that happened before. Behavioural scientists have written books about this wisdom. For us, it suffices to know that if one prevents the first thing happening, then the following thing would not happen. Goethe, in Faust I, says it clearly through Mephistopheles:

"The philosopher enters and proves to you, that it has to be thus. The first is thus, and the second thus, and therefore, the third and the fourth are thus. And if the first and the second were not there, the third and the fourth will never be there."[10]

The conclusion from this is:

> All accidents can be prevented, if the things that would logically result in an accident are removed in time.

These were the two jewels of safety wisdom I promised you at the beginning of this chapter. You have to accept that *most accidents are caused by human acts* and you have to firmly believe that *all accidents can be prevented.* Unless you are convinced of the validity of these two basic statements, you will not have the right frame of mind to support a safety programme in your area of management. The fact that all ac-

9) Ratios for other Himalayan peaks are still worse. Annapurna: 58 fatalities/140 climbers (41%), K2: 60/249 (24%), Kanchenjunga: 40/200 (20%), Dhaulagiri: 56/331 (17%). Source: Der Spiegel, 1/2006.

10) Der Philosoph, der tritt herein, Und beweisst Euch, es müsst' so sein; Das Erst' wär' so, und das Zweite so, Und drum das Dritt' und Vierte so; Und wenn das Erst' und Zweit' nicht wär', Das Dritt' und Viert' wär' nimmermehr.
J. W. v. Goethe, Faust I, Schülerszene

cidents can be prevented is amply proved by those factories that run for years without any recordable accident. In these factories, the works managers are not willing to accept any accident, whatever may be given as its cause. Once you too adopt this attitude, you will see how your people react and do their best to see that factors leading to accidents are discovered well in advance and removed before they end in an accident.

You have to be convinced that accidents do not just happen. They are the final result of a chain of events comprising unsafe acts, supervision failures, unwise management decisions and misled worker attitudes. Only when *you* start believing that all accidents are "prepared" well in advance, will you be able to achieve the one and only goal of safety management, which is: Zero Accidents!

4
Man or Machine?

The conclusion from our experiment in the previous chapter was that the basic causes of nearly all accidents are unsafe acts by human beings and not unsafe conditions of machines. This is not only true for personal injuries, but also for large catastrophic industrial accidents, such as Chernobyl and Bhopal. Look deeper into the reasons why an accident happened, be it the sinking of a ferry ("Herald of Free Enterprise" at Zeebrugge, Belgium), bursting apart of a space shuttle ("Columbia"), a devastating explosion in a chemical plant (Flixborough, England) or just a car crash, and you will always come to the conclusion that some person did something unsafe that resulted in the accident.

Fifty years ago cars did not have crumple zones, telescoping steering wheels, double brake systems, air-bags, ABS, not even safety belts. Today, all cars have these, some cars have air bags on all sides around the passengers. Have the number of accidents reduced over the years enormously? Actually they should be down to zero. But they are not, because the problem is not the car, but the person sitting behind the steering wheel. The person who drives too fast, overtakes in risky situations, or has a drink too many before taking his place behind the wheel.

About a hundred years ago, when machines were in their infancy, there were quite a few mechanical failures due to ignorance of the stresses that a material could bear. Look at all those old bridges using tons and tons of steel to make them stable. Take, for example, the Eiffel Tower in Paris. Did you know that the Tower is not built of steel, but of precast sections of a special type of cast iron called puddle iron, riveted together? Alexandre Gustave Eiffel, who also designed and built the internal supporting structure of the Statue of Liberty, did not trust the steel available at that time (1889) to withstand the stresses of a 300-m high tower, and used over 7000 metric tons of cast iron to

Managing Safety. Kishor Bhagwati

ISBN 3-527-31583-7

build it. A structural engineer told me that today, with the special steels available, it could be built in less than one-fourth of the amount of steel. As a matter of fact, he said, one can even construct it as a concrete column. I doubt whether that would have looked as beautiful and graceful as the Eiffel Tower. Don't ask a Frenchman his opinion on this (and also not on the glass pyramid of the Louvre built by an *American* architect)!

The industrial revolution, started by the steam engine, brought many social and economical benefits, but also posed the danger arising out of harnessing large amounts of power. During the middle of the nineteenth century, boiler explosions were occurring in the US almost at the rate of one every four days, and at the beginning of twentieth century, there were two explosions nearly every day. Most of the explosions occurred in shipping. According to an estimate of the US Commissioner of Patents, in the first half of the nineteenth century, nearly 250 steamboat explosions took place killing about 2500 persons. The worst boiler disaster was of the Mississippi river steamboat "Sultana" in 1865, which killed 1200 people. In those days, most such accidents were dismissed as "Acts of God".

The main reason for the explosions was disregard for the limits of the strength of material used. Operators drove the engines over their limit to get more power from them than what they were designed for. Safety valves were blocked and lead fuses against overheating removed. The steam engine was the harbinger of prosperity beyond imagination, and those few deaths, argued their supporters, were considered to be the price one had to pay for it. It was this argument and the decisions taken on the basis of it by the politicians and owners that resulted in the terrible accidents.

The public uproar at last moved the governments, which till then were reluctant to interfere with private enterprise that had brought all that progress, and introduced safety codes and legislations in the first quarter of the twentieth century. Today, we hardly hear of boilers exploding due to high steam pressures in them. The ones that have exploded in the recent past have been due to leakages and subsequent explosions of the gas that is used to fire them.

We have reached a stage today, where failures due to ignorance of the effects of stress on materials have become negligible. This has led to a relative rise in the percentage of accidents caused by human activity.

An analysis of 550 000 accidents by H. W. Heinrich in 1931 also showed that 88% of accidents were caused by unsafe acts of people. Our experiment has shown that we can confidently state that about 98% of accidents are caused by unsafe acts. Unsafe conditions contributing to the rest of accidents can in most cases be traced back to faulty design or bad workmanship, something that can also be classified as unsafe acts.

When I talk of unsafe acts, I do not mean only acts of workers, but also the decisions of management. The decision-making process is as much a human act as any other activity. I had heard the sentence: "Every accident is a sign of management failure", and I once used this statement at our board meeting. You can imagine the uproar it created. John, the director of our largest unit, a tall, broad and powerful man who had played (American) football for his university in the States, got up and asked me angrily, "You mean to tell me that *I* and my guys are responsible because some *idiot* hurt himself on the shop-floor?" "Yes, John," I answered, "if you select and employ idiots to work for you, isn't that a sign of management failure? Or perhaps the man was not an idiot, but a normally intelligent person selected by your department to work there. In that case, he was not properly trained to work safely. That too is a sign of management failure, isn't it?" He fumed, but did not argue any further.

Let us now go through some major industrial catastrophes and try to analyse why and where things went wrong.

Bhopal, India

The worst industrial disaster in the history of mankind took place in the night of 2 to 3 December 1984 in Bhopal, a town with at that time 800 000 inhabitants in the geographical centre of India. The leakage of a poisonous gas at five minutes past midnight from a factory manufacturing pesticides killed 3000 people in the same night, and out of the about half a million injured, there have been about 15 000 related deaths since. More than 100 000 people are still suffering from various ailments, such as partial or complete blindness, damaged immune system, gastrointestinal disorders, stillbirths, offsprings with genetic defects, etc.

What had happened? About 27 000 kg of methyl isocyanate (MIC), escaped from a storage tank in the factory of a multinational American company called Union Carbide, a company that started off in 1917 in Charleston, West Virginia, as Union Carbide & Carbon Corporation – the name reflecting their earlier history in manufacturing carbon products such as carbons for electric arc street lights and electric furnaces, and calcium carbide for making acetylene for carriage lamps. The heavier-than-air gas MIC spread in the surroundings, killing most people living in a shanty town near the site of the factory. The gas cloud proceeded into the town, surprising people in their sleep, and killing or seriously injuring thousands of them.

Here are the facts. Note particularly the words in italics.

- MIC is a chemical used to make an insecticide called carbaryl.
- The manufacturing process used by Union Carbide is a two-step process. In the first step, MIC is made from methyl amine and phosgene. In the second step, MIC is reacted with alpha-naphthol to produce carbaryl.
- MIC is thus an intermediate product in the manufacture of carbaryl, and not a raw material or a final product.
- Manufacturers often like to store their intermediate products, because if there is some problem in the first step, the second step can continue producing the final product for some time from the stored reserves.
- The whole plant at Bhopal was shut down for major repairs. There was *no necessity of storing MIC*, as the second step was not in operation. In spite of this, *55 tons of MIC were stored in two tanks* designated as E610 (40 tonnes) and E611 (15 tonnes).
- MIC is a highly toxic product with a boiling point of about 38°C. It has, therefore, to be stored under refrigerated conditions, because in Central India temperatures, even in winter, can rise above 30°C. *To save electricity, the refrigeration unit was switched off.*
- Water entered the tank from an empty pipe connected to it. The reason is not known. One assumption is that a worker was asked to clean this pipe with water.
- MIC is known to react very strongly with water. Therefore, if an operation such as washing a pipe connected to an MIC tank is to be carried out, everything must be done to avoid water entering the tank. Normal practice in such cases is to put a round steel plate

(called a slip-plate) between two sections of a pipe that are connected with flanges. Such a slip-plate should have been inserted in a flange between the pipe to be cleaned and the tanks. *This was not done in Bhopal.*

- When water does enter the tank, MIC starts boiling and a high pressure is built up inside the tank. The tanks, therefore, have safety valves, just as in a steam engine. But, as the vapours escaping from the tank are highly poisonous, they cannot be allowed to escape into the atmosphere like steam, but have to be destroyed.
- For this, the venting pipe of the safety valve was connected to a washing tower (scrubber) in which sodium hydroxide solution, which immediately renders MIC harmless, is allowed to continuously trickle down from the top to the bottom, the vent gases entering from the bottom and leaving at the top detoxified. *To save electricity, the pump circulating the sodium hydroxide solution was shut off*, and the escaping MIC could pass the scrubber untreated.
- To destroy any remaining gas after scrubbing, the exit of the scrubber was connected to a flare, which is a sort of large Bunsen burner that always has a pilot flame burning inside. Any gas entering it burns off into harmless atmospheric gases, in this case into carbon dioxide, nitrogen dioxide and water vapour. *The flame in the flare was shut off to save fuel.*
- If a toxic gas leaks out, there is a possibility of erecting a water curtain consisting of a row of vertical water jets to act as a barrier for the further spread of the gas. Such a curtain did exist near the tanks, but because *the water supply was not strong enough, the water jets did not reach the height to build a barrier for the gas moving downwind towards the town.*
- As a last recourse, warning alarms are sounded in the neighbourhood to warn the public to stay indoors and shut their doors and windows. Shanty towns, however, do not have such tight sealing doors or windows. Residents of the town proper could have done this, if the alarm had sounded long enough to wake them from sleep. *But, shortly after the alarms had been activated, they were shut down to avoid causing panic.* Many people sleeping with their windows open died or became blind.

From the beginning till the end, a series of decisions taken by those in charge of factory operations ended in this terrible catastrophe causing thousands of deaths. If all the points *highlighted in italics in the listing* had been avoided, the leakage could have been controlled, and the tragedy would not have happened.

Conclusion: wrong procedural decisions and wilful negligence of basic safety rules by managers in charge were the cause of the Bhopal tragedy, and not unsafe equipment or machines. Had the managers in charge of operations not taken the risks they had, thousands would not have had to die.

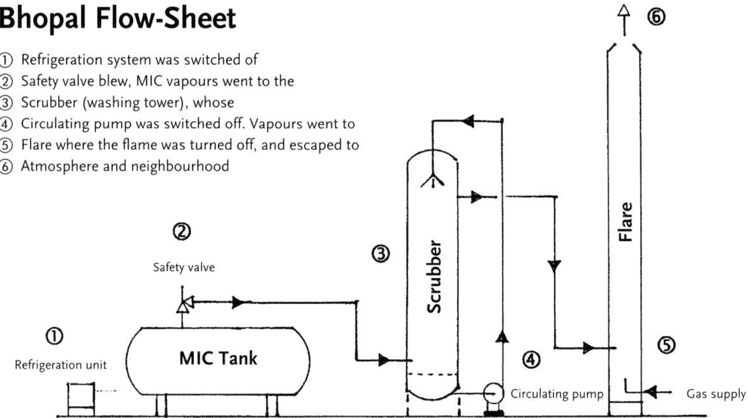

Figure 3 Bhopal flow-sheet

Chernobyl

Early on 26 April 1986, there were explosions in one of the units of a nuclear power plant at Chernobyl, about 100 km north of Kiev in the Ukraine (at that time, still part of the Soviet Union). Although it is said that thousands died from the accident, and hundreds of thousands are suffering from cancer due to the radioactive fallout, in reality, 28 people, most of them engaged in the cleaning of the rubble, died from radiation exposure within four months after the accident. Two fire brigade men lost their lives in action. A further 209 involved with the clean-up were treated for acute radiation poisoning, and out of these, 19 died subsequently from the effects of radiation poisoning. About 4000 cases of thyroid cancer were diagnosed among those who were children and adolescents (0 to 18 years old) at the time of the accident.

Figure 4 The damaged Unit 4 building of the Chernobyl power plant

The survival rate after treatment was nearly 99%. No others in the outside population suffered from acute radiation effects. The fall-out over Western Europe was negligible.[11)]

11) Report of the Chernobyl Forum, 2005. The Chernobyl Forum is an initiative of the IAEA (International Atomic Energy Agency), in cooperation with the WHO (World Health Organisation), UNDP (United Nations Development Programme), FAO (Food and Agricultural Organisation), UNEP (United Nations Environment Programme), UN-OCHA (United Nations Office for the Coordination of Humanitarian Affairs, and UNSCEAR (United Nations Scientific Committee on the Effects of Atomic Radiation).

How could this happen?

The technical details are too complicated to go into here. The following points will show basically what happened.

- In normal power plants, steam is produced by heating water with fuel, such as coal, oil or gas. In nuclear reactors, steam is produced from water by heating it with radioactive elements, which are allowed to have a controlled nuclear reaction, thereby producing intense heat. The steam is used to drive turbines, which are connected to dynamos that produce the electrical power.
- When a plant is to be shut down for maintenance, electrical power is purchased from the power grid fed by other power plants to keep the cooling-water pumps running to prevent overheating of the reactor.
- Management at Chernobyl decided to carry out an experiment to see whether the water pumps for cooling would run long enough after the electrical supply from the reactor was shut down, so that one *need not purchase power from the grid.*
- The reactor design does not permit such experiments. Through built-in automatic mechanisms, deviations from the normal running are immediately corrected, as this would otherwise endanger the reactor's stability. *The management decided to switch off these safety devices to enable them to carry out the experiment.*
- The radioactive elements overheated and ruptured and the resultant force of steam lifted off the cover plate of the reactor, releasing radioactive material into the atmosphere. A second explosion threw up pieces of the burning fuel, allowing air to rush in and causing parts of the reactor consisting of graphite to burst into flames.
- The graphite burned for nine days, releasing the greater part of radioactive material that was the main source of radioactive contamination in the surroundings.
- *Authorities* tried to play down the accident, *did not warn the population* to stay indoors for 24 hours, and did not distribute iodine tablets, which could have prevented the thyroid cancer in the 1800 children.

The Chernobyl catastrophe, therefore, was also the result of a deliberate management decision to ignore strict safety rules and to override normal protective equipment. We once again have to conclude that it was not the machine that caused the damage, but humans.

"The Herald of Free Enterprise"

On the evening of 6 March 1987, a ferry belonging to the Townsend Thorensen Company left the harbour of Zeebrugge in Belgium to take 459 passengers, 81 cars, 3 buses, 47 lorries and 80 crew members to England. Just about 100 metres from the port and after 90 seconds of its beginning the journey, the ferry capsized, bringing death to 193 people.

Figure 5 The *Herald of Free Enterprise* lying on its side in shallow waters off Zeebrugge

The ferry was a so-called RORO ferry, a roll-on roll-off ferry. These ferries have their lowest loading deck very near the water-line to accommodate the maximum number of vehicles. The bow (front) and stern (rear) ends are shut water tight with doors before leaving. Unfortunately in this case, the ferry left the harbour with the bow doors slightly open and water entered the car deck and flooded it, capsizing the ship within no time. A last minute manoeuvre to turn the ferry onto the right avoided the ship sinking in deeper waters, a situation that would have resulted in more deaths.

How could this happen?

The German constructor, who had built the ferry, had provided for an interlock system that would have prevented the ship from moving unless the doors were tightly closed. This was not installed, assuming that the captain would be informed by the person at the door whether the doors were closed or not. Unfortunately, the person who was supposed to inform the captain was sleeping in his cabin, and the captain started off without this information. Not to have the interlock was a managerial decision to save money. After a public inquiry into the sinking in July 1987, Britain's Lord Justice Sheen published a report that castigated Townsend Thoresen, the ship's owners, and identified a "disease of sloppiness" and negligence at every level of the corporation's hierarchy.

I think we now agree that even these big catastrophes are caused by human acts, because taking decisions is a human act too.

Try to go through any other major accident you know, and you will come to the conclusion, that behind all of them was a human being who decided something, either for himself, resulting in his own unsafe act, or for somebody else. The nice proverb "Penny wise and pound foolish" (or "Cent wise and Euro/Dollar foolish") sums up many a decisions of near-sighted managers. The reason behind this is that managers are evaluated by their ability to save money in the short term, and they accept the risk of not spending the pounds, believing that all will go well. They are usually patient when they listen to their engineers and safety specialists asking for more money, but have the feeling that these persons are overdoing it a bit.

I fail to understand why these same managers get weak knees and are helpless when their computer experts come to them and demand new computer systems costing millions every few years and ask that the whole staff be trained in the new software. Is it because they understand abstract computer issues better than concrete safety issues, or because they make their decision out of fear of things they do not comprehend?

5
Why Do Accidents Happen?

Now that it is settled that the main cause of accidents are human acts, let us see how we have been approaching safety till now.

Safety activities at nearly all sites I know concentrate on mechanical safety, i.e. making machines safer, so that people do not injure themselves. Every accident that takes place results in an analysis of the defects of the machine, and more and more safety features are built in to avoid the same accident in future. The analysis hardly ever tries to find the root cause of the accident, because people have not been specifically asked for it (it is not the company culture), are not trained in in-depth incident investigation techniques and, therefore, do not know how to do it. Finding the root cause can be a strenuous job, a good in-depth investigation taking in some cases one or two days. It is much simpler and easier filling out the form for the authorities who do not demand an in-depth investigation and are happy if their formalities are fulfilled, often also internal forms, which follow the same lines.

During my consultation at a large fuel-gas selling company, I discovered an accident reporting form that had, in the section for "Reason of Accident", three choices:

☐ Defective machine
☐ Error of the worker
☐ Residual risk

Guess where most of the crosses were? Nine out of ten accident reports had the cross in the box for residual risk. That was the easiest way out, because any of the other two items would have required some more work. The investigator had done his duty, and had the satisfaction of having finished his work duly.

Managing Safety. Kishor Bhagwati

ISBN 3-527-31583-7

And on what does your company concentrate its efforts when an accident happens? Get the last three accident-investigation reports sent to you (unless you already have them), and check whether recommendations for avoiding the same accident in future include any measures to improve human behaviour, or do they just contain mechanical improvements for making the conditions safer? And if they do mention human error, do they have any suggestion other than asking the worker to "behave more carefully in future"?

Do *you* believe that asking a person to work more carefully in future will make him work more carefully? Wasn't he told that when he started working for your company? How often did our mothers tell us not to do this and that? Did it have any permanent effect? At least, not in my case! If you have grown-up children who go out in the evening (nowadays at times when we were supposed to be back home from our outing), what are the last words you say? "Drive carefully." But their friends are there, they wave back happily to us, and have forgotten what we have said by the time they have turned around the corner.[12)]

Figure 6 Cigarette cartons

12) Nearly every Monday morning, papers report terrible car accidents involving young people who had been to a disco. The report states clearly that alcohol consumption was not the reason. This may be the truth. The reason often is high sound levels in the disco. The extremely high sound levels have an intoxicating effect on people (that being the kick boys and girls seek in a disco). Driving immediately after leaving the disco is similar to drunken driving. So, if you have children who go out with friends to discos, insist that they rest outside the disco for at least ten to fifteen minutes before starting driving. This could save their lives!

Look at all those new drastic warning signs on cigarette boxes in the EU and several other countries, ruining hours of effort and meetings of designers to decide on the art-work of the packaging. More than half of the space is taken up by warning signs. In Switzerland it is still worse, because the warnings have to be written in all the three national languages (German, French and Italian). This leaves hardly any space for the logo!

Do you honestly think this is going to prevent all smokers from smoking? More and more young people smoke today than before. 80% of new smokers are below the age of 18![13] I do hope that the laws and the taxes will help reduce smoking, especially in the young people. The huge warning labels may give the governments the feeling that they are doing something good for the health of the public, but the whole action does not address the real reason why a young person starts smoking in the first place. The root cause of youngsters starting to smoke is not taste, but the glamour of smoking and the feeling of being grown up associated with it, catalysed by peer pressure. Smoking, in spite of all the warnings, is a sign of having grown up, and which young person does not want to seem "cool"?

All technical corrections cost money. No problem with that, if the money spent would definitely prevent a recurrence of the accident. But sadly, it doesn't. You see the same accidents happening again and again. You go on spending more and more money for technical improvements, safety switches and interlocks. But still accidents happen at that place. Are you putting your money in the wrong place? Yes, you are. Have we not discovered that it is the human being who causes accidents and not the unsafe machine? If you would invest the same amount of time and money in improving the work practices of your people, you would have a much greater success.

And why do all those costly technical investments have so little effect? Because people always find a way to bypass the safety switches and interlocks, if they are in the way of ease of operations. In a machine shop of a company I visited, there was a large lathe for trimming flanges over 300 mm (12 inch) in diameter. During operation, the metal cuttings flew around, with most of them falling on the lathe bed. These had to be removed with a long metal hook by the operator from

13) Preamble to the EU Directive 2001/37/EC of 05 June 2001 on the "presentation and sale of tobacco products", Official Journal of the European Communities, L 194/26-34, dated 18 July 2001.

time to time. All moving parts were enclosed in sliding glass doors to avoid the razor sharp metal cuttings, which flew around, hurting the operator. A microswitch was installed that would immediately shut off the lathe if the glass door was slid open. The engineer who had designed this was, I am sure, fully convinced that with this construction of his he had seen to it that there would be no injury to the lathe operator.

When I went past the lathe, the protective door was open, and the lathe running merrily. The microswitch, which should close down the machine if the door was opened, was fixed in a door-closed position with adhesive tape. On being asked, the operator replied that it was too cumbersome to restart the machine every time he had to remove the cuttings. He, therefore, always kept the door open to remove the cuttings when they accumulated too much on the lathe bed. And, after all, nothing had happened to him until now, well nothing, except a nasty cut on the upper arm which, by the way, was covered only half by his T-shirt sleeve, with one of those flying cuttings.

Given time and money, an engineer would design a nice complicated system, using electronics if possible, to counteract an unsafe act. If the only solutions your people can think of are mechanical solutions, you have to be immediately on your guard. The question you have to ask when the money request is presented to you for approval is: "Would this definitely prevent once and for all the same or a similar accident happening again in future?" If the answer is not very positive, ask them to think of something else. We shall be dealing with this in a later chapter, when we discuss incident investigation techniques.

At a printing shop, one of the big metal rollers on an offset printing press had to be cleaned of ink after every printing run. The cleaning was done manually with a solvent soaked cloth rag. To prevent workers from getting their fingers caught and crushed between the rollers while cleaning, the machine was designed such that a push button on the frame of the machine to the left of the roller had to be pushed to move the roller. The worker was trained to first push the button, advance the roller a few centimetres, release the button, clean the exposed area with the rag, remove his hand, and advance the roller again a few degrees to clean the next section. Fail-proof construction, you would say.

But things came out differently. To get his work done quickly, the worker kept the finger of his left hand on the button all the time, keeping the roller running, and cleaned the parts exposed towards him with his rag. The rag got caught, pulling his right hand between the rollers, and crushing his middle and ring finger, which had to be amputated.

The accident investigation team led by the safety specialist came up with a brilliant idca. Now two push buttons, connected in series, were mounted on each side of the frame, so that a person had spread his hands to push both buttons at the same time to move the roller. This way the worker was forced to let go of his rag and remove his hand from the danger zone to push the buttons. This would solve the problem once and for all, thought the team. A few days later I was at that plant, and went past this printing machine. I saw another worker doing the same thing as his colleague before, i.e. continuously pushing the button on the left with his left hand, and cleaning the roller with a rag in his right hand. I went nearer and saw: what do you think? Yes, the other button was jammed with a match-stick in the "on" position.

Engineers – and to that category also belong safety specialists – are trained to discover technical solutions to problems. The more technical the solution, the happier the engineer. They are fully convinced that whatever they build will be operated by the user as prescribed and instructed. After all, they do produce bulky operating manuals, covering any situation that may arise, don't they? But they forget that they are dealing with human beings, and not with machines. The human being tends to forget things, and always tries to find a short-cut if the correct way is too cumbersome or requires more effort. How often have I seen fire-protection doors, which would prevent the spread of fire and should, therefore, be normally closed, that are jammed open with a wooden wedge; or microswitches or safety trip switches that are cellotaped in the open position.

Actually, it is possible to operate a machine which is not 100% safe without causing an accident, if one just takes into account all the deficiencies of the machine.

If you take a taxi in a small town in a developing country, you will be in for many surprises. You would see the driver wearing just flip-flops (or nothing), pushing foot pedals that have no rubber on them. Then he would pull a string from somewhere under the dashboard and start his taxi, steering the taxi with a steering wheel that has a play of about thirty degrees on each side, and finally he would ask you to please hold your door fast, so that you do not fall out. And then he would drive you through the bazaar chock full of people walking on the roads and bring you to your destination without ever hitting anybody. If you ever tried to drive that taxi, you would be hitting a pedestrian every other second. This may not be a very fitting example, but it shows that if one knows the deficiencies of a machine, one can take them into consideration while operating it.

I do not say you should let your workers work with unsafe machines, and tell them to be careful. It is your duty to provide safe workplaces for your workers and issue them the necessary personal protective equipment (PPE). There are methods like Job Safety Analysis (JSA), that dissect each and every action necessary to carry out a job, evaluate whether the action is free of hazards, and if not, take measures to minimise the hazard. Whole processes can be analysed by various such methods, one of the best being Hazop (Hazard & Operability Studies). But, surprisingly, how ever much you invest in using such techniques to make a workplace least hazardous, accidents still happen. Let us be clear on one point: There is no workplace absolutely free of hazards. Not even in the kitchen at our home, or in our garden. If we want to achieve excellence in safety, we have to start working on people, and not only on machines.

First of all, we have to realise and recognise that people just *do not* always do what is expected of them or what you would like them to do. For example the managers who decided not to install an interlock on the ferry Herald of Free Enterprise (indeed!) to save some money, *expected* the captain to always look whether the lamp on the bridge showed whether the doors were closed or not, before moving out of the harbour. This was something he had done hundreds of times, but this one time he didn't, and 193 persons died.

Or take safety belts in cars, for example. In spite of the propaganda and education, many people still did not put on their safety belts when driving. In the US, they tried to install a contact in the belt clip, which would not allow the motor to start unless the clip was closed. So what did people do? They closed the clip behind their back, sat on the belt and drove! Nowadays you have a gentle, pleasing bell that reminds you to put on the belt, not a blaring, loud, unbearable signal. Some even argue that safety belts are redundant now that we have air-bags. That is why a traffic psychology professor in Munich suggested that to really make people drive carefully, one should not install air-bags, but fixed spears to always remind people what could happen to them in case they drove rashly.

Compulsory wearing of safety belts was introduced quite late in Italy, and the police took photographs of drivers not wearing safety belts, so that they could fine them. What do you think was the hottest selling item during the first few months? Shirts and T-shirts with the safety belt printed on them! Every mentality provides its own solutions to restrictions. There is an old adage that says that in the UK everything is allowed except that which is expressly forbidden, in Central Europe everything is forbidden except that which is expressly allowed, in Russia, everything is forbidden, even that which is expressly allowed, and in Italy, everything is allowed, even that which is expressly forbidden. The first time I went to Italy, I was picked up at the airport by a business colleague. He drove over the autostrada at 160 to 170 km/h (100 to 110 mph). So I asked him whether there were speed limits in Italy. His answer, with a typical wave of the hand was: “Yes, we do. But only in theory!”

My experience has shown that there is only one reason why accidents happen:

Accidents happen because people take known risks.

Unsafe acts result from the will to take risks. A person who has hurt himself has done something that he knew could hurt him, but still he did it. Why? Because he had done it a hundred times before, and it had gone well. So he thinks that this time it would go well too. He knows it could be risky, and if, by chance, it did not go well, he could hurt himself. But, after all, his practice has made him perfect, and he is in total control of the situation. This made him take the risk again, but unfortunately, this time it went wrong, and he hurt himself badly. He would like to call it bad luck, but I shall call it overconfidence. His pride in his “mastery” of the situation makes him underestimate the hazard and the risk, and makes him do things he *knows* are unsafe.

Why does a person take risks? Because he is an optimist. If you told a person that the chances of his having an accident with his car this evening on the way home would be one in ten thousand, he would just shrug, and say that that is such a rare chance, that he does not have to worry about it. And then, on the way home, he would stop at a kiosk and make his six crosses on a lotto card. The chance of his having all six numbers right is one in 15 million! But he is confident that his six numbers will definitely come, and he drives home, happily dreaming of his coming trip around the world. This is optimism.

An optimist believes that only good things happen to him and bad things happen to others. He has heard of others having accidents at this particular workplace, but he does not have to worry about it. It will not happen to him, because, after all, he knows the equipment well, and has total control over it. This attitude is not only prevalent at work. Rash driving, mountain climbing, risky skiing, bungee jumping, etc. all live on this attitude.

Men are bigger risk takers than women. What is behind it? One word explains it all: Bravado. Since ancient times men had to provide for their families by hunting wild animals. Physical strength was considered the ideal to achieve, because those who were stronger had greater success in their chase, and could last longer. Taking risks was a necessity in those times, as without risk, there was no food. This type of risk is no longer necessary. Picking groceries from a supermarket shelf and wheeling them to the cash counter does not involve any bravado. But the wish has seemed to remain. That is the reason for taking risks in ordinary life, going in for sports involving the danger of injuring oneself, and watching on TV people who constantly endanger their life by driving like madmen or facing a bunch of hoodlums pointing guns at them.

W. Somerset Maugham wrote a nice short story about some circus artistes, who used to dive into a small shallow basin of water from a great height. The girl, who was to jump, thinks that actually the spectators were waiting for her to miss the basin and die. What they were tense about were the guts she had to face death, and not her artistic talent. What fascinates us when we watch people taking terrible risks are their guts, something that we would never have. But all of us take some risk or other, may it be driving over the crossing when the lights have already turned yellow, or standing on a chair supported on a table when changing a light bulb. Every time we take a risk and have not suffered an accident, we feel we have achieved something great. This love of taking risks, big or small, is the reason we have accidents. The caveman, still lurking within our genes, subconsciously makes us do it. And this is the thing we have to change.

If working safely means not taking risks, then the whole essence of safety training is nothing else but training your people not to take risks. They must be trained to avoid risk. For that, they must recognise risk. Not that they will not take a risk after having recognised it, as we shall see later.

Safety training consists of convincing people not to take risks.

Very often when I discussed with persons who had accidents, why they did a certain risky act, the answer I got was: work pressure. The boss had demanded that a job had to be finished very quickly. To keep the deadline, a shorter route was taken, knowing well that it could be risky, but time was more important. A job had to be done, the production had to continue, and so forth. Let us be clear that many of the accidents that happen are not caused by workers who want to hurt the company. On the contrary: they want to do their best for the company. They are even willing to accept risk and remove a jam from a running machine with their hands, so that it keeps running and the production does not suffer. Punishing such persons is the worst thing we can do. It is the manager's duty to imprint on his workers' minds that no matter how urgent the work is, the manager does not want the worker to take any risk doing it.

The Message from the Manager

I want you to take time to do everything safely, however urgent or important it may be.

Your safety is more important to me than production.

If you cannot do it safely, do not do it.

The manager not only has to say this, and put up sign boards proclaiming this, but he has to act accordingly. Never should he take a person to task because he did not do something that he thought was unsafe, and the production was retarded due to it. And this message has to come from the very top. It should be a part of the company policy, and one of its maxims. All newcomers have to be impressed with this principle, and all the managers have to live it. Only then does one have a chance of climbing up into the first league of safety-conscious companies.

At one such highly safety-conscious company, the stock of cardboard cores (round cylinders on which sheet material is wound, e.g. as in toilet paper rolls) was running out, and the delivery date was approaching fast. So the company called the supplier to immediately send one lorry-load of cores. As all supplier's lorries were on the road, he just ordered one from the market to do the transport. The lorry came to the ramp of the company warehouse to be unloaded.

Now, the problem was, that the company had standardised the height of their ramps in collaboration with all its suppliers. The outside lorry was an older model, and its floor was about 30 cm lower than the ramp. The company's fork-lift driver refused to unload the lorry. The lorry driver argued with him and said he could get a few planks, and roll up and down those planks. This would have been very unsafe, as there was no provision, either on the lorry, or on the ramp, to fasten such planks. So the fork-lift driver said no.

The lorry driver phoned the supplier, who phoned the works manager of the factory. The works manager talked with the fork-lift driver, understood the situation, and asked the supplier to recall his lorry and send one with the standardised height. This took two hours, and the production had to be stopped for one hour because of lack of cores. The fork-lift driver was never scolded for his action because it had resulted in loss of production. As a matter-of-fact, the next company bulletin praised him for his strict adherence to safety and for not taking a risk. This is how really safety-conscious companies work.

Needless to say that this company has an extremely low accident rate, and is one of the most profitable companies in its sector.

What are the conclusions we have arrived at in this chapter?

1. Accidents happen because people take risks.
2. People take risks because they are optimists.
3. Optimists believe that bad things happen only to others, not to them.
4. The essence of safety training is to educate people not to take risks.
5. Nonrisky behaviour must be a company principle, to be followed by all.
6. Management, from top to bottom, must promote nonrisky behaviour.

6
The Underlying Reasons

We have found that human beings often act differently from what one expects of them. You provide a safe workplace, you give them personal protection equipment (PPE), you train them in the techniques required for their work, you show them how to work safely and still, yes in spite of all that, they do something risky (some would say silly) and hurt themselves. Provide interlocks on their machines, and they would bypass them; provide gloves to protect their hands, and they would not use them; provide safety glasses, and they would grind without wearing them. Why, oh why?

The reasons why people do something lie deep within them. It may be in the subconscious, and may not even have any relation to the problem at hand. Let me give you an example.

Some twenty years ago a friend of mine, who was in charge of an analytical laboratory, complained to me that in spite of all his efforts to make the lab assistants (all female) working in his lab wear safety glasses when working, he was far from having any success. He had ordered nice posters saying "Eyes one pair, None to spare", or showing an arm with a yellow bandage with the three black dots on it and asking: "Would you like to join his company?" In his frustration he had decided to start noting down the names of those he found not wearing safety glasses, and was toying with the idea of fining those that were caught the third time. I asked him not to do anything like that, but try to talk with them in a group and find out the real reason why they did not wear safety glasses.

So he called a meeting of his lab assistants, and asked them to tell him why they did not wear their safety glasses. After a lot of here and there, the answer finally came. The safety glasses supplied to them, they said, were ugly, and made them look like old hags (they were really not nice looking – functional, but with no thought to good design whatsoever). This real deep-seated reason had nothing to do with safe-

Managing Safety. Kishor Bhagwati

ISBN 3-527-31583-7

ty, and nothing to do with the conviction that they were necessary. It was just a question of vanity.

My friend then ordered some chic-looking pairs of safety glasses from a different company, put them in the lab, and asked the lab assistants to tell him within two weeks, which types of glasses they would always wear. Not which they liked, nor which were comfortable, but which ones they would always wear. They came back after two weeks, and showed him the pair selected. The frames were stylish red with thin black stripes. These, they said, they would always wear. My friend ordered them, and the ladies kept their word.

So, you see, the reason for not wearing a PPE may not have anything to do with negligence, but may have a deeper reason, completely unrelated to the problem. It is our job to find this reason and undertake something to remedy the problem. The way to do it is by asking questions.

If you see a worker not wearing his PPE, do not shake a finger at him and tell him what he is supposed to do. He would do it sheepishly in front of you, but the moment your back is turned, he would revert to his old ways. It is better to take some time and talk with him about the issue. Ask him, why he does not, say, wear his safety glasses. As a reason he may tell you that they are not comfortable, or that they give him a headache, or that he just forgot them in his locker. If the glasses supplied to him are really not comfortable, ask him to go to the optician and select a pair which he finds comfortable. If they give him a headache, ask him to go to an eye doctor and get his eyes examined, because you are worried about it. If he just forgot them in his locker, ask him to go and fetch them, and do not leave the place till he returns and puts them on before resuming his work.

If you are like me and wear glasses regularly, remember this: for a person who has never worn glasses (and this is the majority of workers), it is quite an odd feeling wearing them. We are quite comfortable with them, although we always use the opportunity to take them off when they are not needed. At least I do, and I have seen many people doing it, especially when one is tired. Well, this is how a person who has never worn glasses feels when he wears them for the first few times. When we got our first glasses, we had the benefit of our sight improving. We could see things more clearly, and that was a great relief. But for the worker who wears them the first time, it is a worsening of his sight. There is something between himself and his objects, and this is the first thing that bothers him. Then there is this odd pressure on the bridge of the nose and behind the ears. Naturally, he feels uncomfortable, and would avoid wearing them as much as he can! So do have understanding for a worker who refuses to wear glasses in spite of the rules, and do not be hard on him.

And, if none of the tricks mentioned above works, then take him aside and convince him why he should wear them. We shall discuss how to do this properly later but let us be clear on one point. In no case can you accept his not wearing safety glasses where these are necessary, and where the danger of eye damage exists. You should not leave his workplace, just reprimanding him to wear glasses "in future", because, if a flying metal piece hits his eye and he loses one eye shortly after you left, then it was you who were negligent in your duty, and not only he. You "permitted" him to break a safety rule by letting him work in a hazardous situation without wearing glasses. Thus, in this case, the underlying reason behind the accident was *your* negligence of your duties, and not only his unsafe behaviour.

What is the message a manager sends if he goes past an unsafe act without remedying it? "I accept your unsafe behaviour." Such negligence, if practiced often, creates the impression in the minds of the workers that the manager does not really care whether one works safely, or not. The manager makes the proper noises on his safety rounds, but one can forget them once he has left. He is not very serious about it anyway! Risky behaviour then becomes the norm, and accidents happen. And the manager, who has documented his safety rounds, has the big excuse of having done all he could. The accidents that have happened are just bad luck. Kismet! he calls it.

Through his negligence to cultivate a climate of non-risk behaviour in his department, the manager supplies a major underlying reason for accidents to happen.

Such underlying reason works in the subconscious. It is what we call *habit*. The OED defines habit as a "settled tendency or practice". For a thing to settle, it must go on for some time. Practice is the way things are done customarily. Both senses mean that an act must be carried out repeatedly to become a habit. A one-time action is not called a habit. We saw in a previous chapter that accidents happen to people who have done the unsafe act hundreds of times without hurting themselves. And then something does not go the same way it did until then, and the accident happens. Acting in an unsafe manner was the habit of a person who hurts himself.

Habits can be good and bad. There are good habits like waiting for a machine to stop running before putting one's hand in it, and bad habits like not waiting for it to stop completely before putting the hand in it. The first leads to safety, the second, to accidents. If we want to prevent accidents, we have to cultivate good habits, and weed out the bad habits. This requires some work. Bad habits are difficult to remove, especially when they make life a bit easier. Comfort is the keyword of our present society. Look at all the advertisements. They all promise to make life easier for you, whether it be when operating a washing machine, or preparing food or shaving in the morning. Invent a product that requires less effort than its predecessor, and you are sure to have a product that sells well. Even payment for the product can be done in "easy" instalments!

It is not impossible to cure bad habits. It does not require superhuman effort. Many companies have had success in transforming bad habits into good habits. You can do it too. You just have to learn the proper way of doing it. A few chapters later you will see how it is done. The most important thing, however, is that you must *want* to do it. You must believe that an accident-free factory is possible. You must have an intense desire of seeing to it that your workers do not get hurt as long as they are under your care. You, as a manager, can achieve it much more easily than any safety officer. The only thing that is required is your visible commitment.

What exactly is this much-maligned word "commitment"? It is a stool with three legs. The legs are: Interest, Involvement and Investment. If any of the three legs is missing, the stool will fall down. In a later chapter we shall be discussing in detail what exactly the three legs mean. If the word Investment has put you off, do not worry. I am not talking of millions of euros. I am just talking of your investment in terms of time. In the field of safety, the returns on this investment are tremendous, both morally and financially.

7
How to Make People not Take Risks

No human being likes to be constantly told by someone else what he should do. Any person with some self-respect revolts at the idea of being dictated in his actions. This is the reason why so many rules and regulations do not work. Our history is full of people who have sacrificed their life to gain freedom for themselves or for their country, like Jeanne d'Arc, Mahatma Gandhi or Nelson Mandela, or people who have heightened the self-respect of their people by making their country great (even if it meant robbing the freedom of others), like Alexander the Great and Napoléon. The US Declaration of Independence, in its beautiful English, proclaims that all men are created equal, and endowed by their Creator with certain unalienable Rights, that among these are Life, Liberty, and the pursuit of Happiness. The electrifying speech of Patrick Henry that closed with: "Give me Liberty, or give me Death" has modelled our attitude towards freedom and independence. We honour and respect the men who said these things, because they showed us how important it is to be free and independent. We even idolise them, and build monuments glorifying them.

Those who have grown up believing that freedom of thought and action is everybody's birthright, cannot be asked to blindly follow some rules they do not agree with. One can use force and the threat of punishment, like the authorities do. Some managers I know believe in this method, and do sometimes have marginal success, but the accident rate, as also the absenteeism rate in their departments, not surprisingly, is high. They have to continually increase the pressure to set their goals through. And the worker morale in these departments is pitiful. Everybody grumbles and complains about the methods of their manager, who himself is deaf to all such complaints, because he never takes time to hear them, and, if he does hear them, waves them off as just normal griping.

Managing Safety. Kishor Bhagwati

ISBN 3-527-31583-7

People will only do by themselves that of which they are convinced. All other external forces, to which people may bend like palms in a storm, will be of temporary nature. The moment the force is gone, they will upright themselves, like the palms after the storm has passed, and carry on doing things the way they were doing them before. This is why inspections, carried out in an authoritative manner, have such little permanent effect.

How then can we achieve a permanent change in people's behaviour?

What is the driving force behind a person's behaviour? It is his self-interest and his inner feeling. Nothing drives a person more than these. Behind every great achievement is the self-interest of the achiever. He may believe he is being altruistic, and doing things only for others, but deep within him is the self-interest in fulfilling his ego of being altruistic that drives him. Take any example, and analyse it in depth, and you will discover this truth.

Take the Olympic sportsman who makes superhuman efforts to break the existing record, and becomes a world champion in his discipline. He trains for hours, does not enjoy life like others in his age group, even does illegal things like taking doping drugs to achieve his goal of being the best in the world. Did he do it for his country, so that it has one more medal? Did he do it for his family, friends or his company? No, he did it all for his own sake. He did it for the reward, which he rightly deserves for his efforts.

There are even examples where people overstrain themselves without wanting to become the best in the world. Take the many marathons that take place nowadays in nearly every city that has a few thousand euros or dollars to spare. That the very first marathon runner Phillipides, who ran about 42 km[14)] from the battle plains of Marathon to Athens to announce the victory of the Greek forces against the invading Persians during the Greco-Persian war in 490

14) The current official marathon distance of 26 miles and 385 yards (42.195 km) was established purely by accident at the 1908 Olympic Games in London (The official venue for the 1906 Olympics was Rome, but this was cancelled due to the volcanic eruption of Mount Vesuvius, and instead London was chosen.) The course was originally laid out to be 26 miles long from Windsor Castle to the finish line in the White City Stadium. However, it was then decided to add the 385 yards so that the race would finish at the Royal Box. In 1924, the Olympic Committee formally adopted the distance as official.

BC fell dead of exertion at the end of his run after shouting "Rejoice, we conquer!", does not bother anybody. The jogging and running boom of the 1960s led to a marathon boom in the 1970s, and has become today a test of fitness for amateur athletes. Apart from the happiness aroused by the hormone serotonin that is produced during long-term strain, it is the sense of having reached the goal and having received the badge that satisfies the inner ego.

A person's feelings drive him even further. The love in the family, of a man for his wife, or of parents for their children, is one such thing. In the working world it is the feeling towards his superior. Unfortunately, not all managers are born leaders whom the people will follow willingly to please him. But what nature has not given him, he can learn if he digests the basics of right attitude towards his workers and acts accordingly. These basics can be condensed into one phrase: Respect your workers and their ideas and feelings.

Let us accept it as a basic fact that people act willingly through inner motivation, and not through external force. And this is how safety management is to be viewed. You may be able to use the whip to bring the workers "in line" with your ideas, but they will hate you for it. Some managers do not want to be liked. They will never have people going through fire for them, they would drive the workers in the hands of the unions, tensions would grow and finally, work, i.e. production would suffer.

I knew a plant manager in a metal rolling mill I was consulting, who used to rule with an iron hand. For him, the workers were just trouble-makers, who had to be kept strictly in line to ensure production. He never had an open ear for any grievances and used punishment as motivation. He told me once that he runs his plant like a military general, expecting strict obedience to his orders. Nonobedience would not be tolerated in his plant. He walked through the plant to find who was working unsafely, and sneaked up on him to catch him unawares *in flagranti*.

I talked alone to some of the workers in his plant. They had not a single good word to say about the manager. The moment someone spotted him entering the plant, he and others would take long detours to avoid meeting him. The manager thought he was doing the right thing by strictly controlling everything, because he did not trust his workers. And there was reason for him to think that way. The productivity of his plant was really not optimum, small mistakes that people did ruined whole batches of products. And the accident rate was highest in his plant. He had three times as many accidents as other departments at the site. His frustration at this made him more aggressive. And then, a fatal accident happened! A worker fell into a cauldron of molten metal and vaporised instantly.

But that did not change his methods. He gave all the blame to the poor dead man, accusing him more or less of having brought trouble to his plant. His orders became stricter, his rounds through the plant more seldom, and his interest in safety decreased, instead of increasing, because he now had to spend all his time to fulfil the production demands, which were lagging. Within a year, another person died at work in his plant. The top management at last realised the unsuitability of that person as a manager, and he was dismissed. If the company would have had a system to recognise the safety commitment and involvement of a plant manager earlier, two lives could have been saved. We shall see later how evaluation of the managers' involvement in safety can be continuously monitored by the top management.

The only way to make people act safely is by addressing their inner feelings and values. A person convinced of the fact that any unsafe act he does *could* go wrong and hurt him, would think twice before doing it. Once you recognise this, then you have found the key to solving the safety problem. You just have to learn to use the key properly. This proper use of the key, however, requires a different mind-set from the one you may have today. Just as democracy is not a state of government, but a state of mind, you have to be willing to let a safety-oriented mind-set be your innermost conviction. You have to put freedom from injuries of your workers on the same level of importance as production, sales and profits. If you are not willing to do this, then just forget the whole thing. You will have no chance of achieving an interruption-free production and the high productivity that goes with it.

Putting safety on the same level as other managerial values means that you will never let the interests of production, sales or profits override the necessities of safety. Not that safety concerns should be higher than other values, but at least not lower! If *you* do not give this importance to safety, your workers also will not.

Safety becomes only as important as the managers make it.

Many managers love to talk of their commitment to the safety of their workers, calling the workers the company's most valuable assets. They write such sentences in their Safety Policies (or Environment, Health and Safety (EHS, or SHE, whichever way you prefer) Policies), which are nicely framed and hung in the reception hall, all meeting rooms and on all notice boards. But unless this commitment is visible to the workers and felt by them, such words are not worth the paper they are printed on.

For example, I hold the arm of my son tightly, wave my finger at him and tell him. "Listen, I love you. I can give it to you in writing. If you want, with the seal of a notary public on it. So you better believe it." Would he believe it? Not unless I have shown my love by holding him in my arms, drying his tears when he was hurt, and have been kind to him when he had a problem. It is the same with commitment to safety. Should the workers believe you, just because you have put the words on the notice boards, but they have never felt that you have really treated them as the company's most valuable assets? Why all this hypocrisy?

You must, first of all, be convinced that you have selected good and intelligent workers, and that these workers are capable of giving their best, and *will* give their best, if treated properly. Unless you have respect for the person and intelligence of your workers, they too will have no respect for your person or intelligence.

Secondly, you have to recognise that an intelligent person like your worker also has good ideas, and you can tap this source for better safety. Just because you have studied and have become managers, does not mean that you have better ideas and solutions. As a matter-of-fact, the best solutions are produced there where the problems are, and not in the office of the manager.

If you respect the intelligence of your workers, you should recognise that you can convince him better through logical arguments, than by just giving orders from above. You have to know where his interest lies and what his values are. If you do not, you have to find out. Normally, his interest lies in his own well-being and in the happiness of his family.

Once, when crossing a road in a factory I was at, I saw a driver of a supplier's lorry driving without a safety belt. I requested him to stop, and asked him, whether he had a family. Yes, he said, he had a wife and three children. Then I asked him, whether he loved them. Of course, they were his everything. So, I went on, how would they feel if he were seriously hurt in an accident, probably losing an arm or even life because he had not put on the safety belt? It would be unimaginable grief, he said, not only for him, but also for his beloved family. Then I asked him, would he like something like that to happen to him, and on his shaking his head in denial, asked him further how he could avoid the situation. He smiled at me and said, "I get your point. Naturally, by wearing the safety belt all the time." I thanked him for his time and waved him good-bye.

About six months later, I was at the same factory, again crossing the same road, when a lorry stopped and blew his horn to attract my attention, and the driver waved his hand asking me to come near the cabin. At first I did not recognise him, but when I came nearer, I did. It was the same driver I had talked to six months ago. He just put his thumb through the safety belt he was wearing, raised it to show it to me, and said, "Never without!" He was smiling from ear to ear, a smile I would not so easily forget, and we shook hands, both very happy at the change in attitude that had taken place.

If, instead, I had just stopped him, told him that company rules required that he must wear a safety belt, and threatened to black-list him if I saw him not wearing a belt a second time, the effect would not have been the same. Within the factory area he might have sometimes worn the belt, at other times not, hoping that nobody would catch him, and outside the factory just taken it off, because he had not found it necessary before, and did not find it necessary now to follow some (silly) restrictive rule.

The way should be now clear to you. If you want people not to take risks, you have to convince them that taking risks does not pay, because they are the ones who would suffer most. If you see a worker not wearing gloves, ask him who would suffer most if he lost a finger? Does he think it would be his superior, or the works manager? Does he think the company would stop and go bankrupt just because he lost his finger? No, non, nein, nyet! The biggest sufferer would be himself. He and his family would be the only ones who will be short-changed. You do not ask him to wear gloves. You do not point out any regulations to him. You also do not order him to do anything. You just help him think over the consequences of his action, and leave him alone to take the necessary steps to protect himself from the consequences. This is the technique we shall be talking about in the next chapter.

"It is a fact of human make-up that people cannot be commanded in attitudinal matters. People cannot be commanded to change their attitudes any more than they can be commanded to hope or to believe or to like something."[15)]

15) Thomas R Krause, "The Behaviour-Based Safety Process", John Wiley, 1997.

8
The Myths of Management

Managing is a hard job. The higher you get, the more strenuous it gets. Decisions have to be taken constantly, often on matters one does not have much knowledge about. One does not have time, even if one wanted, to go deep into the subject, and take the wise decision. And sometimes, going deep does not help much either. The matter may be so completely out of one's area of education or knowledge that one has to take other means of deciding. This is the managerial intuition.

Prof. Cyril Northcote Parkinson, who formulated Parkinson's Law that states that "Work expands so as to fill the time available for its completion" in his famous book "Parkinson's Law" (London, John Murray, 1958), gives an excellent example of this dilemma of managers. A company wants to build an atomic reactor with an estimated cost of £10 million[16]. The board of eleven members meets to decide on this. Of the eleven, four, including the Chairman, do not know what an atomic reactor is. Of the remainder, three do not know what it is for. Of those who know its purpose, only two have the least idea what it should cost, one Mr Isaacson, and one Mr Brickworth.

Mr Isaacson objects to the choice of the consultant and the contractors, and suggests taking somebody else like Prof. Levy as consultant and Messrs David and Goliath as contractors. The Chairman objects, arguing that the proceedings are already quite advanced, and there is no need for a new advisor or contractor.

Mr Brickworth is almost the only man there who knows what he is talking about. He has quite a few questions in his head, but where should he begin? None of the others would understand what he was saying, and he would have to begin by first explaining what an atomic reactor is, and he already knows that the others would never admit

16) Mid-1950s figure.

Managing Safety. Kishor Bhagwati

ISBN 3-527-31583-7

their ignorance. Better to say nothing. So he decides not to make any comments.

The Chairman signs the contract on behalf of all board members. Total time for this item involving £10 million: 2½ minutes. The next item on the agenda is the approval of a bicycle shed for the use of the clerical staff costing £350. This item takes forty-five minutes. Why? Because everybody can visualise a bicycle shed and the sum of £350, and can argue about it extensively. Prof. Parkinson concludes that "The time spent on any item of the agenda will be in inverse proportion to the sum involved."

I highly recommend his books to all managers, because it shows how the real world is and works. They are very wise books and at the same time entertaining. I suggest the book "The Law", first published by Murray in 1979, which is a collection of all his laws in one volume.

A manager has to live with all these imponderables and still take decisions that sound wise and correct. Lacking concrete facts to base his decisions on, he starts relying on myths. Myths, which sound very good, and look great, but have inherent fallacies in them resulting in wrong decisions. Let us take one such myth.

Safety Management Myth Number 1
Policies printed on paper and exhibited everywhere will change people.

It is easy to write safety policies. Take someone who can write well, probably from the HR department, give him samples of policies from other companies to write his text, get the text checked by the legal department, give it to the internal or an external publicity organisation to make a nice layout with pictures, get it signed by the works manager and/or the CEO, see that it is printed in nice colours on heavy paper, and hang it in nice frames at various spots in the company. Done!

The policy looks nearly like a religious declaration of what the company values most, such as the environment and the health and safety of its employees. Some of them even promise to put these values on the same level as commercial values. Complicated long-winded sentences make it sound like something very official – and therefore important. The word "commitment" sprouts out of every paragraph. It is a document for the shareholders and visitors, who could not care less.

My experience, at least, has shown that it does not reach those for whom it is meant, namely the employees.

Figure 7 Dilbert cartoon

Whenever I visit a factory, I go to the shop floor and talk to the workers. I have often asked them whether they have read the safety policy of the company, and what they think of it. Not a single worker has yet told me that he has read and understood the policy, that he agrees with it, and that he finds it true and honest. The answer I mostly get is: It is just a piece of paper. None of our mangers act according to the policy.

If you do not believe what I wrote in the last para, just take half-an-hour's time, go to the shop-floor, and ask a few workers to describe to you what is written in the policy and what it means. You can even start this right outside your room in the office building. But read it yourself before doing any of this. It could be that you yourself are not fully aware of what is written there. And do not be disappointed at the answers you get.

So why write and exhibit such a policy, if it does not serve its purpose? Because the corporate world expects it of you. All major companies write such policies, and you have to keep up with the Joneses. It is supposed to be good public relations too. The outside world should know how seriously you take important issues like environmental protection, so that they change their views of looking at the industry as the destroyer of the world, as told to them by the Greens. People do recognise the benefits the industry brings them in the form of jobs and goods, but their opinion of the moral values of a company is not high.

A Policy does not convince and change people, unless it is accompanied by corresponding acts. The policy should be the final reference point of all decisions. If any issue arises that can be solved in two ways, then the way offered by the policy should be the one that is to be taken. So that this can be done, the wording in the policy has to be short, succinct and without double meanings. Learned persons, especially lawyers, are not capable of expressing themselves in a generally understandable language, especially one that is understood by less-educated workers. It should be written in a language that the workers use. This can be achieved very simply by letting a small committee composed only of workers rewrite the policy in their language! We shall be talking more about this when we talk of communications later.

Safety Management Myth Number 2
The Safety Department should manage safety.

This is the most widespread belief. And it is a false belief. The safety department is an essential management tool that can be used by managers. The safety people are there to coordinate site-wide programmes and give advice when the managers require it. The Safety Department does not *make* safety, as little as your doctor makes your health. Your doctor can advise you on how you should live to take proper care of your health, but what he says is to be done by *you*. If he recommends that you take a walk of at least twenty minutes everyday, it does not help if he goes for the walk during the time you are munching snacks in front of the TV. It is you who has to go out, come rain, come shine, and walk all of those twenty minutes.

To manage something, you have to have power, without which your word has no authority. What powers have you given your safety professional? Can he go out in the plant and stop a machine he feels is highly hazardous to the worker working on it? Can he tell a worker to leave the plant immediately because he is doing some horseplay that may hurt another person? Can he tell a contractor to stop working on an urgent project because he is not wearing the necessary safety equipment? No, he cannot, because he is not bound in the line management, but is outside of it as a staff member. He has no say in the operational side of the factory. If he wants any of the above things to be done, he has to approach and convince the responsible plant manager.

However, his is a very important function. He is the storehouse of information in the field of safety, he knows all the relevant laws and regulations, he visits fairs to inform himself on the latest developments in the field of safety, he represents your company in various safety-related committees of your industry federation and he has a network of contacts in the industry that can help you solve your problems without your company having to reinvent the wheel. Above all, he is a very valuable advisor for you and your managers. But that is all he is. An advisor – not a doer. So do not put the load of improving safety on your safety professionals. Then on whom? In the next chapter we shall discuss who is actually responsible for the safety of the people.

Safety Management Myth Number 3
Safety Department should best report to Human Resources.

Quite logical, isn't it? Especially now that we know that most of the accidents are caused by humans and not by machines, what better place for the safety professional to report than to the human resources (HR) manager? With this arrangement the works manager also has a filter between himself and the safety department that can sort out which demands of the safety people are really important and worth considering. HR managers are supposed to be able to separate the wheat from the chaff; hence they can reduce the time you may have to spend at management meetings by presenting the case only for really important projects.

The safety specialist has thus first to convince the HR manager, who normally has not risen from the production side and in most cases has no engineering knowledge. Also, as he does not bear the responsibility for the operational side of the site with all its intertwined units that contribute towards production of goods, sales and profits, his view is limited to personnel problems and union matters. Something that may be affecting a part of the production in an indirect way is not immediately apparent to him. His decision, therefore, is based purely on intuition and financial considerations.

The head of the safety department of a plastic film making plant described to me his conversation with his superior, who was the site HR manager. It went like this:

Safety manager (SM): Mr Headcount, we have this problem in the packaging department where we have quite a few accidents, with people cutting themselves in the fingers, in the arms or in thighs with the cutting knives we have. I have been recommended a new type of knife that could avoid this, and I think we should purchase it.

HR manager (HM): How many knives are you talking about, and what do they cost per piece, Mr Spendthrift?

SM: We would need initially about 200 pieces to equip all our workers in packaging with a knife. Counting on a renewal rate of one knife per worker every two months, about 1000 additional knives would be required this year. The knives are special spring-loaded knives that retract their blade automatically into the grips once the cutting is finished. They cost about 10 euros apiece.

HM: Let me see, that would be 12 000 euros every year! That is a hell of a lot of money just for penknives. Is that justified?

SM: Sir, we have two to three accidents with deep cuts every month, and about 10 smaller cuts. These knives would reduce the accidents substantially.

HM: But can't you find cheaper knives? Knives are just knives, and the knives we are issuing at present are not bad, are they? The people should be more careful when cutting.

SM: We have tried everything, but the knives are pressed down when the cutting is being done, and once it leaves the material, it has too much force on it to stop immediately. They then slide further and produce wounds in the workers. This knife, as I said, withdraws its blade automatically once the cutting has ended, and thus it cannot hurt anybody.

HM: Well, it's all only theory at present. I suggest you purchase, say, ten knives and try them out. And if then the accident rate does not fall, we forget the whole thing.

SM: But, Mr Headcount, this will not give us a representative value. We must buy at least a hundred to see any significant effect.

HM: Mr Spendthrift, I think I have made myself quite clear. You try out these knives, and then we shall see. By the way, where do the knives come from?

SM: It's the German company "Schneidgut".

HM: Ah, well. Germans do make good knives. But see to it that they are from Solingen, where the best German knives are made. I always look for the sign of Solingen when I buy a German knife, and have been very satisfied with the quality.

The HR Manager sits back fully satisfied with himself because he has carried his management responsibility properly and helped his safety man to take the right decision. The safety professional goes away dejected and at a loss how he should proceed, and fed up with trying to convince someone who has never worked in a plant and has no idea of production hazards.

There is an important issue here that needs to be looked at and remedied. It was the safety professional who discovered the problem with the knives, and not the manager of the packaging department. The manager did not see it as his responsibility to think about it and ask the safety person for advice. He believed it was the duty of the safety department to cure the problem. If it were he, who had asked for a solution, fully knowing how much production stoppage a deep cut results in, then there would have been no financial problem.

The safety professional should always directly report to the works manager, who has an overview of all the various factors affecting the productivity in a plant. The works manager would be in a much better position to evaluate whether 12 000 euros are worth the reduction of cutting accidents by 50 or more per cent, or not. Apart from that, the direct reporting of safety to the same position where all the plant managers also have to report raises the status of the safety department, and with that the importance of safety at the top level. Being a direct reportee, the safety manager also takes part in management meetings, and could bring in ideas at a decisive level. Remember what we said earlier? Safety will only be as important in the plant and with the workers as the top manager makes it. We shall be talking later about effective safety organisational structures.

Safety Management Myth Number 4

Compliance with laws and regulations is all we have to do in the field of safety.

Anyone who believes this, invites regulators to come up with more and more regulations, asks them to control the compliance of all these regulations, and awaits fines in case of noncompliance. Do you really want it this way? Do you want to employ an extra person who just goes around and checks whether all the various regulations are being complied with? Believe me, it is a full-time job!

I once consulted a railway company, and had the opportunity of riding in the locomotive with the engine driver. (It was a wonderful consulting contract, fulfilling the dream of my childhood dreams. I could sit next to the engine driver, breeze along the landscape and see the rails through the front windshield. The greatest of thrills was entering a famous London station with its 24 tracks, passing through a maze of signals and switches and at last coming to rest at the buffers.) I asked the driver whether he had a guidebook or an operating manual. He nodded and passed on to me a loose-leaf binder in A5 size[17)], about 12 cm thick. I asked him whether he knew the book thoroughly. He shook his head and said: "It's no use, because every time an accident happens somewhere, I get 20 more pages full of instructions. I just can't read and master all that, especially not in the legal language it is written."

The following two points deserve consideration:

Firstly, regulations always represent a compromise between interested parties. They are the least common denominators. They also have to consider the poor performers and cannot drive these to bankruptcy through investments they need to make to meet very strict limits. Regulatory demand is the minimum one should do. They are not the last

17) Paper sizes in Europe and most other parts of the world (except North America) are based on the original 1922 German Standard (DIN 476) on paper sizes, which prescribes that the ratio of the shorter side of a paper to its longer side must be $1:\sqrt{2}$ i.e. 1:1.41. This has the advantage that upon every folding of the paper in half along the longer side, the ratio of the sides remains the same, half of the previous shorter side now being the longer side ($\sqrt{2}/2 = \sqrt{2}/(\sqrt{2} \times \sqrt{2}) = 1/\sqrt{2}$, i.e. $1: \sqrt{2}$). Starting with A0, each halving produces a higher A number. Thus, normal letter paper is A4 size, because it is obtained by folding the basic size, A0, four times. A0 was defined as having an area of 1 square metre. Envelopes have the same proportions, but are a little larger (B sizes). One can, therefore, fold paper of any size as often as one wants, and it will fit into a smaller envelope. An A4 sheet will fit into a B4 size envelope, folded once it will fit into a B5 size envelope, or folded twice into a B6 size envelope. This standard is now also the international standard ISO 216.

and the final wisdom, and just sticking to them does not lead to excellence in any field.

I was once visiting a factory employing about 800 people near Hamburg. It was a well-run factory, and had a canteen on the first floor of a building, where lunch was served. The canteen could sit about 200 persons at a time. Attached to the canteen and open to the seating area was the kitchen, where food was prepared and served. The canteen had its own separate stairwell, and at the bottom of the stairs was a door to outside, which was the only exit. Thus in case of fire in the kitchen, people would have to rush down the staircase and exit the building via the only door. This door, I noticed, opened towards the inside!

It is well known that if a fire breaks out, people start panicking and want to leave the building immediately. At such times, a door that opens inwards is pushed shut by the pressure of the people wanting to get out, and the bigger the panic, the greater the rush. Such doors must always open towards the outside.

I asked the manager who was with me, why the door was designed this way. His answer was that a door opening towards outside would require a small roof to protect the door from weather and rusting. To save the money this roof would have cost, it was decided to let the door open towards the inside. But, I said, isn't it more sensible to let the door open the other way? His answer was that there was no regulation in that state demanding that doors of rooms where people gather, like the canteen, must open towards the outside, and that they were thus fully in compliance with the regulations. He had argued the same way with the fire inspector, who had to give in and certify the situation as complying with the law.

Secondly, laws and regulations can only regulate machines, and not humans. Regulations are absolutely necessary to render a workplace safer by demanding safety devices, endurance tests and safer chemicals. There still are many companies out there who would not even provide personal protective equipment to their workers, if the laws did not demand them. You can see that clearly in developing countries, where labour is cheap, there is no worker compensation and a person earns per day less than what a pair of gloves would cost. This means that regulations, to be effective, must be enforced. Enforcement means frequent audits by authorities, and fines for noncompliance. Nobody wants to be at the receiving end of such systems.

We have recognised that accidents are caused by humans and their unsafe and risky actions. No regulation or law will prevent them taking risks. Relying solely on regulations will not get you where you want to be. You *have* to follow and comply with them, but you should never stop there. It is not the end-point, but the beginning. Complying with regulations is like cutting the vegetables and the meat and arranging the different herbs and spices on the table. Going further than compliance is like cooking the dish and finally enjoying it. How you rate as a safety chef depends on what you have achieved: whether you rate one, two or three Michelin stars.

There are many other such myths prevalent at the managerial level, such as “Spending money will cure all safety problems” or “Punishing people who had accidents will improve the worker morale”, etc. Think over them by yourself. And give some thoughts to what other myths exist in your daily managerial life, and write them down just for fun. It will be a most enlightening experience.

9
Who "Makes" Safety?

It is now time to decide who does what in the field of safety. Printed below is a list of some safety-related work that is normal at a production site. Go through the list and make a tick against the item you think should be the safety professional's job. Those you do not tick fall in the area of line managers. We shall discuss each item individually and see what the right answers should be to create conditions which will lead to excellence in safety.

What is the Safety Professional responsible for?

1. Improving the attitude of workers towards safety
2. Keeping the managers' head free of all safety-related matters
3. Discovering the need for safety training programmes of individuals
4. Organising site-wide safety seminars on selected topics
5. Writing the Company Safety Policy
6. Checking compliance with the Company Safety Policy
7. Bearing responsibility for site safety performance
8. Investigating all accidents at site and writing investigation reports
9. Coordinating safety activities of the site
10. Chairing the Central Safety Committee
11. Participating in and coordinating outside safety audits, e.g. corporate audits, certification audits, etc.
12. Being the information source for safety-related legislation and appliances
13. Being the advisor to management on safety issues

Managing Safety. Kishor Bhagwati

ISBN 3-527-31583-7

1. Improving the attitude of workers towards safety

Safety management does not mean putting the burden of working safely on the shoulders of the workers, and the managers worrying about "more important things." Either the attitude towards safety changes in the whole company and at all levels, or there is no chance of progress. Changing attitudes is not easy, and posters do not help much. It is true that attitude drives one's behaviour, but modern safety psychology has discovered that the way to change attitude is through a change in behaviour. "A change in behaviour leads to a change in attitude"[18)]

Safety must become a part of the company culture, company culture being "the set of shared attitudes, values, goals and practices that characterises a company or corporation"[19)] How can a (mere) safety professional produce such a radical change in the company culture? The change can only be brought about by those who have the power to bring about major changes in a company, and these are the leaders, the most powerful among them being the topmost leader, i.e. the works manager. If you had put your tick here, it would have been at the wrong place.

2. Keeping the managers' heads free of all safety matters

Would you believe that this is the job description of a safety professional that I have heard from many managers when I asked them what they expected of their safety professionals? One works manager told me very plainly that he had no time to worry about safety, and he had therefore told his safety officer never to bother him with safety matters. "What else have I employed him for?" he asked me. If the interest in safety at the top is that low, then the company can forget having high-class safety and the higher productivity and better profits that go with it. The only persons who can relieve the managers from their "burden" of safety are they themselves. Done properly, it is not a burden at all. Actually, believe it or not, it can be fun. If you follow the suggestions in this book, you will find how satisfying good safety is and

18) Thomas R Krause, The Behavior-Based Safety Process, John Wiley 1997.

19) Merriam-Webster's Dictionary.

how much pleasure it gives one to have an accident-free site. No tick here either.

3 Discovering the need for safety training programmes of individuals

Who knows about the where, the when and the how much of the knowledge gap of a worker? Who can define what additional training his people require? Training sessions mean time away from production. Hence one wants training to be short, but effective. To be this, it must be narrowly defined. This definition can only be done by those who have frequent contact with the workers, know them and see where a worker needs additional training. In short, they are his immediate superiors. The manager receives this information from his supervisors (if he asks them!), formulates the requirement, and asks the safety department to arrange for a training session. Knowing the production situation well, he can also schedule the sessions better. In short, the definition and administration of the training is the plant manager's responsibility, the safety department providing the sessions with either internal or external resources. No tick here please.

4. Organising site-wide safety seminars on selected topics

Safety seminars are different from training sessions. They are general in nature and cover a variety of subjects in the field of safety. For example, a seminar on safety management on the lines of this book would be a type of seminar for all management levels and categories. When I give these seminars, I insist that financial, purchasing and personnel managers also take part in it, because they too have to contribute to, and be part of the culture change the company wants to introduce. Other seminars could be, for example, on new legislation, or on hearing protection, etc. Such global seminars are necessary if the company wants to achieve a certain goal by means of a concerted action, and this is best organised by someone who has a site-wide point of view. Organising such site- or company-wide seminars, therefore, falls in the area of responsibility of the safety department. They have to evaluate the market by speaking with their network contacts, select the best consultants, and coordinate and administer the seminars. This item, therefore, definitely deserves your tick

5. Writing the Company Safety Policy

We already talked about the company policy in last chapter, and shall be discussing it in more detail later. Suffice it to say that the company policy is a statement of guiding principles of a company, and this can be written only by those who establish these principles and believe in using them as their guiding principles, i.e. the top management. If a safety officer were to write it, he would write what and how it should be. The management will then start trimming it here and there, finding the wishes and demands too excessive till what remains is the least the management would like to commit itself to. However, if a subgroup of the management committee wrote it, it would automatically receive the approval of the board. And, knowing that everybody knows who the authors are, it may even result in a document that is more stringent than the one the safety officer would have produced. Note, however, the question of the language of the policy mentioned in last chapter. We shall be discussing it later in the chapter on communications. No tick here either.

6. Checking for compliance with Site Safety Policy

In the last chapter we mentioned that a safety policy should be as short and succinct as possible. If done in this sense, it would not have more than eight to ten short sentences. They are principles of operation and not instructions for operating. One very important statement in the policy would be: "Managers are responsible for safety". How would the safety professional go around checking this statement? By asking how the managers see this? Or by asking workers whether the manager really takes this responsibility? And what does he then do with his findings? Present them to the works manager and perhaps make enemies all over the place? No, checking for compliance with the policy needs interpretation, and this is best left to those who have to interpret the policy, namely the managers. A tick here is again not justified.

A safety officer can check mechanical safety systems, personal protective equipment, fire-protection systems, or legal compliance. His expertise is required when purchasing new equipment to check their mechanical and operational safety.

7. Bearing responsibility for site safety performance

Many managers believe that this is the main job of the safety professional. This is why they have employed him. They wanted to reduce accidents, because they were higher than their industry's average, and they did not want to be seen as poor performers in this field. So they searched for a good safety professional and ordered him to quickly reduce the accident figure. They put him in the HR department, so that he has support from the HR manager, because safety people are not regarded as being high in the hierarchy of such a company, and they did not want to be bothered by him too often. But what authority did they give him? Hardly any. At least not enough to make changes in the minds of the managers, where the problem resides.

One basic principle of human management is:

> No responsibility without corresponding authority.

The only persons who should and must bear the responsibility for safety performance are the managers at all levels, each one individually responsible for his area of management. If the organisation is properly tuned, they not only bear the responsibility, but are also accountable for its shortcomings. We shall see later how this can be put into practice.

8. Investigating all accidents at site and writing investigation reports

We shall be talking about accident and incident investigations and the methodology of carrying them out in detail in a separate chapter. But here I just would like to correct a common practice in nearly all companies I have consulted. If an accident happens, the plant person immediately calls the safety department and expects someone from that department to take over the investigation and write the investigation report. Of course, the plant people are there to help with any detail required, but the rudder is given over to the safety specialist. After all, he is an expert, isn't he? And it's his job, isn't it? The answer to this is: Expert in safety matters, "Yes". But it being his job, "No".

Let us consider this: Who is responsible for giving the worker his work? His manager. Who is responsible for approving his vacation? His manager. Who recommends him for his pay rise? His manager. Who is responsible for his safety? The safety department!?! Isn't there something illogical here? One cannot be responsible for all the work-related aspects of a person, but not for his well-being and his freedom from injury. It is a part of the whole basket of management responsibilities, and if you are caring for all other aspects and neglecting the safety aspect, then how good do you consider yourself to be a manager?

The person who leads the accident investigation team and writes or gets the report written is the victim's immediate superior, or someone next higher in the line management, but not the safety professional. It should be someone who knows the victim well, knows his work well, knows how he works, has perhaps also some knowledge of the victim's personal problems and habits, etc. This can only be his immediate superior, and not someone from the safety department, who has perhaps never met or talked to the victim before. No tick here either.

9. Coordinating safety activities of the site

This is truly the safety professional's job. His field is the whole factory, and he has to see to it that all the various departments in the factory have a more or less uniform approach to safety. The safety department has to collect safety statistics, discover trends in injuries, see which department has frequent problems in a particular safety field, give that department feedback on the trend, and help managers improve their safety. I said "help managers", not "do it for managers". The initiative for this must come from the manager himself. Safety department just feeds him the data, and perhaps suggests how to solve a particular problem. Being on the watch for new legal developments, he can inform the site management on coming things that would affect the site as a whole, or a certain production unit, and recommend activities that need to be started in time to be ready when the law or the regulation comes into effect. So the tick here is absolutely right.

10. Chairing the Central Safety Committee

Members of the Central Safety Committee should come from all departments on site. For manufacturing plants this includes the site manager, the plant managers, and senior members of the other staff functions (HR, financial, etc.). The Departmental and Plant Safety Committees, which we shall discuss later, also has workers and works council representatives as members. The Central Safety Committee can be led only by one person, and that person is the site manager. The head of the safety department, which we have seen as the coordinating department, can only act as the secretary. He can arrange the schedules and the venue of the meetings (which take place usually once a month), and coordinate the agenda with the works manager. He can write the minutes of the meeting too, and give a presentation on the safety statistics of the site. But lead he cannot, and should not. Your tick should not have been here.

11. Participating in and coordinating outside safety audits

Outside safety audits, for example corporate safety audits or audits required for safety management certification, e.g. OHSAS 18001 certification, are site-wide activities that need to be coordinated. OHSAS is the Occupational Health and Safety Assessment Standard published in April 1999 by the British Standard Institute in association with other national standard and certification bodies. It is compatible with ISO[20] standards on environment (ISO 14001) and quality

20) ISO is short for the International Organisation of Standardisation with its office in Geneva. ISO is not a jumbled up acronym that should have been IOS. It is derived from the Greek word ***isos***, which means equal. We find isos as a prefix iso- in words like isotherm or isobar (lines on maps connecting places of equal temperature or pressure). The name has the advantage of being valid in each of the three official languages of the organisation – English, French and Russian. For example, the English acronym IOS would not have been valid for the French designation *Organisation internationale de normalisation*.

(ISO 9001). For this type of coordinating work, the safety department is ideally suited. Some of these audits take a whole week, and someone has to prepare all the necessary documentation before the visit of the auditors, and be ready to supply any additional information they may require during the audit proper. So the tick goes to the Safety Department.

12. Being the information source for safety-related legislation and appliances

Manufacturing plants cannot be expected to have all the information on latest legislative issues or newest safety appliances. The safety department is the library for all such information, and should be in a position to answer questions from the plants on these matters. The safety professional has to update his knowledge continually through professional magazines and visits to trade fairs.

The safety professional is also your site's eyes and ears to the outside world, collecting all the relevant information on safety aspects of operations and informing you in time to take action to comply with a new regulation before it comes into effect. He does this by joining working groups and attending meetings of your trade federation, and participating in technical conferences. In this way he also builds his network of colleagues from other companies. For example a bill is introduced in the parliament to reduce noise levels at workplace. You would not read about this in your newspaper at this stage. Your trade federation monitors all such activities (if it is a well-functioning and active federation) and informs your representative, i.e. your safety department with copies of the drafts.[21)]

Your safety officer gets this advance information and starts a site-wide programme for getting noise levels at workplaces – known for their high sound levels – measured by an accredited institute. He collects and analyses the data and invites noise-control experts (whose names and references he can obtain from his network friends) to propose noise-control measures for reducing the noise levels up to or be-

21) In certain countries like Germany, the government has to send the drafts to the federations of affected parties, who have the right to give their comments before the draft is finalised. The EU has such a system too, albeit not as formalised as in Germany.

low the new legal levels. He presents his project to reduce sound levels to the management, manages the project and finally gets the results again measured by the accredited institute. This is the type of job a safety officer has to do. Your tick here would have been correct.

13. Being the advisor to management on safety issues

Here at last is what the safety professional is really there for. He is your in-house consultant whom you can entrust all your safety problems, and he can work out a solution for you. Exploit his experience and knowledge when deciding on things concerning safety at your site. Let him be your "Minister for Safety Affairs". Once he gets this trust from you, you will see how he changes his method of working. Instead of being a general factotum for plant managers, he rises through your repositioning in his self-esteem and gets the self-confidence to think more as a manager than as a dogsbody. I hope you had made your tick here.

Appendix 2 lists what a job description for a safety professional should contain. After having rewritten his present job description, see to it that it is communicated to all managers at the site.

Conclusion

I guess you have realised by now that safety professionals cannot manage safety of a site, because they are limited in their authority. To make them responsible for safety, without giving them the authority to intervene in production matters is ridiculous and unfair. He is the brain behind the safety effort, but not its hands and feet. Who, then, should take care of safety?

The answer is very clear: The same persons who take care of various other factory-related matters, such as production, raw materials availability, logistics, cost control, people allotment, etc. These are the line managers and the senior staff. The line starts right at the top and ends at one layer above the lowest. It is the responsibility of all line managers, and they are to be held accountable for it.

10
Management Tools

Did the last chapter contain a few surprises for you? I do hope so. Do you see a tsunami of new work rolling towards you? Has your inner resistance already started building up to oppose any change that may entail more work for you? Do not worry. An old proverb says: "A craftsman is only as good as his tools." So what you need now are the proper tools to fulfil your responsibility without undue stress. And these tools do exist. Used properly, they will make your work of managing safety easy, and even pleasurable.

Figure 8 Four pillars of safety. (Courtesy: The British Museum, London.)

There are four pillars on which the temple of Supreme Safety stands. They are:

Managing Safety. Kishor Bhagwati

ISBN 3-527-31583-7

1. Visible and perceptible management commitment
2. Safety visits by managers
3. Worker involvement and authority
4. Thorough accident investigations

The first two of the pillars are your job, and the other two are things you can influence by simple means. But just as the temple needs all the four pillars, so also the goal of supreme safety, which requires all of the above items. You neglect one, and the building starts getting unstable, you neglect two, and it crumbles altogether. One more simile can be derived from this building: all the four pillars have to be equally strong, and have to be maintained at that strength at all times. For example, you plan to carry out safety visits and you have decided on and published a certain frequency at which you will carry them out. If the frequency starts dropping, i.e. you always find some other more important appointment at the scheduled time, or you just don't feel like going, this aspect of the safety management starts getting weaker and your success, which you may have achieved till then, starts getting fissures.

One can also compare it with flying a plane. Let us assume you are the pilot. You give full power for the take-off, and after the take-off go on climbing till you reach your cruising altitude. At this point you can ease the throttle back some to maintain the altitude, but you can never afford to turn the fuel off, otherwise you will crash. In terms of safety, the cruising altitude is reached when you have zero accidents for at least three consecutive years. Till then, you have to maintain the flow of the fuel at a constant rate. We said you are the pilot, and that is what you truly are in your area of management, may it be a complex site, or a small plant. Just as the pilot is responsible for flying the plane, and not flight attendants or passengers, you are responsible for the success of the company's safety efforts.

In the following chapters we shall be discussing each of the above points individually. These four are the basics of safety management. There are many other points we shall be discussing, but without these four being in place, all other efforts will not be crowned with success.

Just as we postulated in the last chapter that no responsibility should be put on the head of a person without giving him equivalent authority, management wisdom demands:

No authority without accountability

Responsibility within an organisation must be agreed upon between the person giving the responsibility, and the person accepting it. Did you notice the word "accepting"? Yes, the person on whom the responsibility has to rest must know exactly what he would be responsible for, whether it is within his area of competence, and whether the time frame given to him is reasonable from his point of view. In short, he must accept the responsibility put on him with his eyes open. Let us take a simple example.

You are a supervisor in the public works department of your town, and you want Mr Spazzare, the sweeper who is working for you, to see to it that a certain area of the pedestrian zone looks scrupulously clean every day. So you give him this responsibility. Mr Spazzare regards the area he has to keep clean, and judges that he cannot clean this large area in one day. He mentions this, but you are of the opinion that it is not as large as he complains, and any person who is fast enough should be able to do it. Of course, you yourself have perhaps never taken a broom in your hand or, if you have, never tried to clean anything larger than your courtyard, especially not an area as large as the one you intend to give to Mr Spazzare. So your demand is not based on experience, but just on your gut feeling. Mr Spazzare, however, whose job is cleaning large areas, has judged that he cannot clean it in one day.

Being the boss you do not deter from your demands, and warn him that you would inspect it every day to see whether the work has been completed. Do you think you would have success? Do you think you have improved the work relationship with Mr Spazzare, and have motivated him to work more? What do you believe he thinks of you? The same as what I hear when making rounds in factories, interviewing workers and asking them what they think of their superiors who pride themselves on being hard taskmasters.

And Mr Spazzare will not be able to fulfil the demand that is put on him. You will always be "disappointed" in his work, and the tension between you will grow day by day. He does not feel obliged to do what is demanded of him, because he knows it is not possible. He also does not put in any extra effort, because he knows you are not right, and that you will not find anybody else who can do it. He finds your demand excessive, so why should he break his back doing some additional work for someone who cares so little for him and his views? All that you have done has not made him like you. That's it!

If, however, you had asked his opinion before deciding on the area you wanted, you would have reached a reasonable solution. If you had asked him how large an area he can manage in one day, he would have told you. His answer could have been half the area. If it was not enough in your opinion, you could have asked him for two-thirds of the area. You both finally agree that the work should be finished in one-and-a-half day. The main thing is that it has been *arrived at mutually, and is acceptable to both*. Now you can put the responsibility on his head, and he would bear it, and fulfil your wish willingly. And if the work does not get finished as promised by him, you can take him to task. But not for the work *you* envisaged, but for the work *he* committed himself to do. This too, is a very important rule of fair management:

Never decide on anything that subordinates should do without involving them in the decision process and getting their agreement.

Coming back to our previous statement on accountability, anybody whom you grant authority to do something must feel responsible for its results and must be made accountable for his performance. If you give the responsibility for injury-free operations to your managers, then they must know that if they do not carry it out properly, they will have to suffer. I know many companies where the responsibility for safety is given to the managers "on paper", but nothing happens to them in case they do not take care of the safety of their people and the accident rates in their department go on rising, as long as they meet their production targets.

Put safety targets, agreed upon by both of you, on an equal level with production targets, and failure to meet either of them should result in the same negative effects for the manager.

Let us now take in the following chapters the four pillars one-by-one.

11
Pillar 1: Total Management Commitment

The word commitment is a favourite of all writers of policies, mission statements and advertisements. Wherever you look, the word hits you squarely in the eye. Companies selling things are committed to give you the best quality, service companies are committed to give you high-class service, chemical companies are committed to keeping the environment clean, and railways are committed to running their trains punctually.

For me commitment is a three-legged stool standing on legs I call the three *i*s. They are: **I**nterest, **I**nvolvement and **I**nvestment.

Figure 9 The Commitment Stool
Leg 1: **I**nterest Leg 2: **I**nvolvement Leg 3: **I**nvestment

1. Interest

Firstly, you have to be *interested* in safety. Interest does not require deep knowledge. You do not have to be a safety specialist to be interested in safety. The results of excellence in safety are: freedom from injury for your workers, better overall productivity, and improvement of financial figures. If you are interested in these three, you should be

Managing Safety. Kishor Bhagwati

ISBN 3-527-31583-7

automatically interested in safety. Once you are interested in a thing, you would like to know more about it, and you will be willing to spend some time getting to know it better. It is like your liking the taste of good wines. Once you get interested in them, you start informing yourself about the various appellations, grapes and vintages. The more you know about wines, the higher the pleasure of choosing and degusting the wines. And one day, perhaps, you start dreaming of a trip to the various châteaux of Bordeaux, travelling from Médoc over Paulliac, St. Julien, Haut-Médoc and Listrac to Margaux, and celebrating the wines there on the spot.

This is how interest works. You have to start getting interested in safety. Not in the technicalities, but in its injury-prevention potential. If, deep in your heart, you do not consider the workers as mere figures who have to do their work and to keep production running, but have some place in your heart for their feelings, their likes and dislikes, and their desire to reach home safely, then you are on the right track. On top of this, if you knew that you really can influence their safety, so that they can go back to their families in the same condition as they came to work for you, you are already there. The fallout of their injury-free working, such as better productivity and quality should not be your primary aim. Just take care of them, and they will take care of your profits.

Take care of your workers, and they will take care of your profits!

The hope of this book is that after reading it and experiencing the joys of achievements in safety, your interest will turn into a passion, a passion that will drive you to ever higher performance.

2. Involvement

The word interest is derived from the Latin words *inter* = between / among, and *esse*[22]= to be. Literally, therefore, interest means being among or between something. This means that a person who is interested is always among the thing he is interested in. The root of the

22) *est* is the third person singular present of *esse*.

word involvement is not much different, it is only stronger. It comes from *in* and *volvere*[23] = to be wrapped in. The involved person is thus totally wrapped in the thing he is involved in. This meaning of the word "involve" shows us clearly what is desired. If one is interested in something, one is also involved in it, and one does not consider it to be something outside his sphere of being. Interest can be awakened, but involvement is to be exercised. For example if you are involved in a matter, you will not neglect it, but will work actively towards its success.

In terms of safety management this means not staying outside of safety matters, but being actively interested in safety projects, safety progress and accidents. Your people must know that *you want to be informed* of all such things, and that you always have time to listen to them. You want to know of every accident that happens at your site, you want to know how the victim is, you want to know about the progress of accident investigation and you want to know what countermeasures have been taken. For a major accident you want your departmental or plant manager to come to you and report to you personally on these things. Your involvement is the proof of your interest.

I am told that in an international company known for its safety performance, the preliminary report of any accident that happens anywhere in the world must be on the CEO's desk within twenty-four hours. He always reads it and reacts to it! (In many companies that have copied this reporting system, thereby expecting it to improve safety magically, the report usually goes unread and lands either in some file that the secretary keeps, or finds its way into the wastepaper basket.) If it has been a serious injury, all employees get an e-mail from him expressing his sorrow and an appeal to see to it that they are more careful in their work.

After reading the report, it is said that he immediately picks up the phone, and asks either the works manager or the country manager the following three questions:

1. How is the victim? Does he have the best medical attention our company can afford?[24]

23) From *volvere* we also have the word *volva* that meant womb in olden times.

24) This question is not that relevant in the industrialised countries with their social net, but in developing countries, where there are no social insurances and hardly any medical insurance, this question is vital.

2. How did the accident happen? and
3. What have you done to see to it that the same or a similar accident *never ever* happens again?

And if the person at the other end of the phone cannot answer these questions properly, or the CEO is not satisfied with the answers, that person is ordered to come to the headquarters within three days and give the answers to the CEO personally. The first question is repeated by the CEO periodically till the victim is back at work, and then, during his next visit to that site, he takes time to go and talk to him.

3. Investment

If you are interested in something and are involved in it, you do not mind investing in it. Both time, and money. If you play golf, this is obvious to you. Did you count the amount of time you spent reducing your handicap or perfecting your drive? And what about those fancy irons you spent more than 150 euros on? All hobbies need investment in terms of time and money. Here we are going further than investing for our personal pleasure or benefit. Your investment in safety is going to benefit many others, benefits beyond your imagination.

The word invest comes from *in-* and the Latin *vestire* = to clothe. Let us use it in this sense. We have put safety in our hearts (interest), and we have identified ourselves with its fate (involve). Now let us clothe it with investment.

What does this investment consist of? Only a little bit of your time. The common complaint is that one does not have time for such unimportant things as safety. It is a specialist's job, and the specialists should worry about it. How many of you have studied financial accounting, or human psychology? Do you leave all financial matters in the hands of your financial expert and all personnel matters in the hands of your HR professional because both are specialists? Don't you sit sometimes for hours together with these specialists analysing situations and giving instructions, although you may have studied engineering or chemistry? The previous CEO of BASF, one of world's largest chemical companies, Bernhard Timm had studied mathematics and astronomy. If the fate of your factory is put in your hands, you just cannot afford to leave safety matters aside, but take them as seriously as money and personnel issues. More so, now that you have seen that bad safety can undo many other cost-saving efforts, such as rationalisation, streamlining, etc.

How much time are we talking of? A total of only two to three hours a *month* – no more! I am talking of concentrated time periods, and not something you shove in fragmentarily between other important matters. Just as you close your door and tell your secretary not to let anybody disturb you when you are talking with your financial or personnel manager, so too you must put only one thing in your mind when you are occupied with safety. For example, if you are on one of your safety visits, leave your mobile phone in your office, and tell your secretary that you do not want to be interrupted during your visit.

Take time for safety

These two to three hours a month are required for your safety visits, your participation in the Central Safety Committee meetings, your conclave with the safety officer and your prodding for answers in safety matters. Is that asking for too much for an increase in quality and productivity? Three hours a month for all these good things, not to mention a harmonious relation with the work force, an increase in worker morale and a minimum of problems with the unions.

In a company I had consulted, the works manager took his safety responsibility very seriously. He had scheduled his regular safety visit to the plant at 10 a.m. Suddenly he got a call from the corporate sales manager that a very important customer wanted to visit him, and that he would bring him to the plant at 10 a.m. The works manager told the sales manager that he had scheduled a plant safety visit at that time, and would the customer please come an hour later? This was nearly unheard of in that organisation before. A customer, especially a big and important one, had always to have priority over anything else, and his wish had to be a command. The salesman was incensed, threatening to call in the CEO, but the works manager remained steadfast. The customer visit had to be rescheduled at 11 a.m.

When the customer came, the works manager excused himself for delaying the visit. To the surprise of all present, the customer said he was very much impressed by the importance the works manager gave to safety in his plant, and he felt reassured that with an attitude like that, he had confidence that he would get his goods always punctually and with a high quality. And the company got that big order. Later, the customer confided to the sales manager that it was this importance to safety that the works manager had given that had finally decided for him who should get the order.

12
Pillar 2: Safety Visits – The Basics

The best way to demonstrate your commitment to safety is by going on safety visits. One could, of course, call them audits if we use the word in its proper sense. The root of the word audit is the Latin verb *audire*, which means to hear. That is exactly what you should do when you go on your round of a safety visit: listen to what the workers have to say, and not vice versa. But the word audit has acquired a different meaning since the 17th century[25]. In larger organisations, it has got a negative flavour, as it is used for detailed inspections by corporate inspectors or by official bodies. This is the reason I have chosen the word visit. A visit is usually something friendly, and therefore agreeable and pleasant. (When it is not, the English language has the word "visitation".) A visit is also caring, like a doctor's visit to a patient's bed when he makes his rounds. In no case is it a trip to catch people *in flagranti* doing wrong things, and to punish them. It is not, and should not be allowed to become an inspection.

A safety visit is an opportunity for you to talk with your workers in a relaxed atmosphere about safety matters and what they think and feel about it. It also enables you to convince workers not to take risks because of the negative effects risks will have on them, and helps motivate workers to work risk-free. It also shows workers that you do care for their safety enough to spend time talking with them about it, and that you are truly committed to what you say in your policy of them being valuable assets. The key message is: you do practice what you preach.

Other managerial advantages you will have are: you will really get to know your plant properly; you will be able to explain to people what you expect of them; most importantly, however, you will have the opportunity to know what they expect of their management. Talking with

25) More about audits in Appendix 5.

Managing Safety. Kishor Bhagwati

ISBN 3-527-31583-7

them will help you recognise where the weakness in safety management lies, so that you can cure and remedy situations that are potential safety holes.

In former Eastern (Communist) Germany, the percentage of women working as plant operators and mechanics was, and is, much higher than in the West. At one such site near the Polish border, the works manager and I went on a safety visit to one of the plants as a part of his training in carrying out safety visits. We saw a woman mechanic repairing a pump without wearing her safety gloves. So the manager asked her why. She explained that she had small hands and that no safety gloves were available for her size. All the gloves she had been issued just slipped from her hand as they were too large, and because of it she had no secure hold on the work piece. So he asked her whether she had approached her supervisor or the purchasing people for this problem. Of course she had, she said, she had very often asked her supervisor and also directly contacted the purchasing department, but the answer was always that smaller sizes did not exist.

Not fully convinced of this, the manager, after returning to his office, called the purchasing department and put the question to them. The answer was: the catalogue of the company supplying the safety gloves listed size 8 as the smallest size, but the worker in question required size 6. Thus it was clear to the purchaser that smaller sizes of safety gloves did not exist. He had already informed the supervisor of the worker about this more than a year ago, and also the woman worker. The works manager asked the purchaser to come to his office with the catalogue. It turned out, that the catalogue sizes did stop at size 8, but the catalogue was for men's gloves! The manager immediately phoned up the manufacturer of gloves, who confirmed that they did have smaller sizes, and that they would immediately send a size 6 and a size 7 to try out which was the better fit.

This small incident showed the manager three weaknesses in his organisation. One: the woman's supervisor did not follow-up the matter as diligently as the manager had done, although it was only a matter of a few minutes, and he let the woman work unsafely, putting her in danger of an injury. Two: the purchaser did not see it as a worrisome situation, and did not call the manufacturer, or search the market for some other manufacturers who could supply smaller size gloves. He did not care whether the woman had safety gloves or not. Three: the hierarchical system diluted the complaint of the woman to "just bickering unnecessarily" and did not take her seriously. Thus in just one visit the manager gained an insight into the working of his company, something he would have never gained otherwise. He knew now what needed to be changed to close such "safety holes".

Question the validity of any statement that seems illogical to you.

Such bureaucratic trenches within the company are quite common. They create situations that jeopardise safety. They are mostly between staff functions and line functions. Staff considers itself separate from "the working population", sometimes even superior to them because of their white collar, and in no way responsible for line problems like safety. This includes even the safety department, if it has been created as a bureaucratic staff department, just managing the paper work concerning safety. I have an example of this.

In a seminar for supervisors I gave in Luxembourg, one of the participants, a maintenance-shop supervisor, complained that he did care for his workers, but nobody in the corporate office seemed to. He described the situation as follows: one of his workers had prescription safety glasses made by a local optician, who was the sole appointed supplier for such glasses for the company. One day, six months ago, the glasses fell down and one of the lenses broke. This showed that the lenses were not made of unbreakable material and thus not suitable for use as safety glasses. So he wrote to the safety department immediately, describing his experience, asking that the matter be examined by the safety department, and requesting information on which other optician the worker could go to get proper safety glasses. He got no reply. So after a month, he wrote a reminder. Again no reply. He had been repeating this now since six months, and waiting all the time for a reply, which was not forthcoming.

I asked him what would he have done if his son had been hurt in an accident, he had gone with him to a hospital and had been made to wait hours, in spite of constant appeals, before somebody attended to the boy. He said he would have made a hell of a lot of noise and banged on the table to get somebody to take care of his child. Why then, I asked, did he not go to the safety department, banged there on the table after he had not received an answer for a week, and demand a solution to the problem? After all he was responsible for the safety of his people, and somebody would have lost an eye if the glasses were not proper safety glasses. His wish to see that his workers were not exposed to hazards should have been stronger than his sticking to proper channels. Next day before the seminar started, he informed me that he had done what I had suggested, and had received the right information immediately.

That solved his problem, but not the problem of the system. I admit that this happened in a government organisation, where red tape and bureaucracy reign supreme. But a little bit of this mentality exists everywhere, where safety is considered the job of the safety department. The powerless safety department sinks into bureaucratic lethargy to protect itself from receiving blame for things they cannot move, and concentrates on producing lengthy reports and complicated charts that are not understandable, hoping to impress people with their diligence and thoroughness. What a waste of talent that can be used with gain for the company if someone had really understood their real function and mission and the job description would have been coined with that in mind? As already mentioned, a job description of a safety professional is given in Appendix 2.

How can safety visits contribute to lowering the accident figure? Because they help us nip the accident in the bud. We have seen that an accident just does not happen by itself, but has been "prepared" long in advance. Behind every accident there is an unsafe act that was carried out hundreds of times but did not end up as an accident. These unsafe acts go unnoticed, if they are not observed consciously. They continue being carried out till, one day, the accident happens, and they come into the limelight when an accident investigation is done. With the accident investigation we can at most prevent a recurrence of the same or a similar accident, but not the accident that has happened.

Accidents, like the visible costs we talked about in Chapter 3, are also only the tip of an iceberg. The tip represents all the accidents from first-aid cases to fatalities. If you see only the tip and are not aware of the mass below the tip, you will never understand why the tip is there. The tip is supported by the mass below, and as long as the mass below the water level is there, the tip will always be there. The mass below, in our case, are those many instances of unsafe acts. (See Figure 10.)

The iceberg has one more property. If you cut off a portion of the tip of the iceberg, what will happen? The iceberg will swim up again

Figure 10 The Accident Iceberg

to compensate for it. You go on cutting more and more of the tip, but it does not disappear, because the hidden mass always compensates for it. Unless you start reducing the hidden mass, the tip will always be there. What is above the water line is a logical result of what is below. To remove the tip, therefore, we have to start working below the water line. This is where the unsafe acts lie.

Safety visits discover such unsafe acts and stop them by convincing the worker why he should not carry them out. This is nipping the accident in the bud. Unless we work at this "underwater" level, the chances of reducing accidents are meagre, because unsafe acts, carried out a sufficient number of times, will automatically result in an accident.

During safety visits, one also discovers unsafe conditions. We have already realised that even unsafe conditions are the results of foregone unsafe acts. However, a safety visit does not try to analyse unsafe conditions. A safety visit gives us a snap-shot of the situation, and we register both safe acts and unsafe conditions. Both are noted down in the report, together with the planned countermeasures to remedy the situation. The quality of a safety visit is reflected in the ratio of acts to conditions discovered. The ratio of acts to conditions can also be used as a measure of the quality of the safety visit. The greater the ratio, the better the safety visit.

Before we go into the techniques of safety visits, the question is: who should make these visits? The answer is: *all* managers, line managers, as well as those in staff functions like finance, HR, IT, etc. If the purchasing manager in the example with safety gloves had been integrated in the safety effort of the site, he would not have let the woman work without safety gloves, but would have moved heaven and earth, including getting special size gloves custom-manufactured to reduce the hazard of injury for people with small hands. Because carrying out safety visits does not require any special knowledge but only common sense, anybody can learn it. I have taught it to over three hundred people at all levels and in all sectors of the industry. What one needs are open eyes and some powers of observation.

Just introducing a system does not mean that it will run smoothly. Managers need instruments to tell them whether the ideas they have introduced and want to be followed are really being followed or not. I have known quite a few companies that started off on their path to superior safety very enthusiastically, but the top manager, not having enough time to check whether his managers were carrying out their duty of safety visits as prescribed, could not plug the loopholes some of his managers were using to avoid doing their duty. The system started corroding and lost its drive gradually till what remained was lip service by the managers and the company.

To avoid this, it is necessary that strict and regular checks are carried out to see whether the train is still on the track. This is done by a Safety Visit Check Table that is reproduced in Appendix 6. This table is put up by the safety department, and it shows whether the managers have fulfilled their obligation or not. The safety department being on the distribution list of all the safety visits being carried out on a site knows exactly which manager has done his duty punctually. The safety professional presents this table once every quarter at the management meeting, and those who have not yet fulfilled their quota have to explain to their superior why they have not, for example, carried out their safety visit, say, since the last three months.

The table also shows the quality of the safety visits carried out by using evaluation numbers we shall be discussing later.

This type of control necessitates that the superior himself has done his duty and carried out his safety visits as noted in the safety visit table prepared in advance by the safety department with the agreement of all concerned. You cannot demand from your people what you yourself are not willing to follow. Being a role model is a difficult job, as any father can confirm who tries to teach his children manners.

I once carried out a safety visit training with the financial director of a petroleum company. We travelled together in his chauffeured limousine to the refinery. The works manager and certain other high officials of the refinery were waiting at the gate to receive him. Now, if you want to enter the plants in a refinery, where highly volatile and explosive petroleum products are handled in tonnes, you have to take great care. You have to wear an overall of flame-resistant material, safety boots, a safety helmet, safety glasses and carry with you an escape mask. Then you have to empty your pockets of lighters or matches. Finally, before entering, you have to watch safety instructions on a video that explains what hazards are there, and how to behave in case of an emergency. After you have filled out a questionnaire that checks whether you have understood what was in the video and have answered all questions correctly, you are allowed to pass through the gate.

The works people wanted to spare all such trouble to the board member, so they had planned a route for him through offices and a mechanical workshop, for which he did not require the full safety clothing. But *en route* to the factory, the director and I had discussed the training, and I had insisted that we go into the plants, where the music actually is. He was agreeable to it. So he told the works manager that he would like to visit the plants, and wanted to be equipped with the safety clothing and other equipment. You can imagine the consternation in the faces of the refinery people, but they had to concede to his wish. So we both dressed up, watched the video together, filled out the questionnaire and then crossed the fence-line. He and I went to the plants alone, as I had insisted, accompanied by a guide who drove us to the right places. Before we planned the trip, he had some theoretical lessons from me on what to do on a safety visit. And he did it very well. He talked to people, asked them the right questions, showed his honest interest in knowing what the workers did, and never showed off his position.

When parting from the refinery, he was asked by the works manager, who had spent two uncomfortable hours in his office worrying about how the "unplanned" visit must have gone, how the financial director had liked it. The answer was: he liked it so much, that he had decided to come back to the refinery after three months to carry out a similar visit of the plants he had not visited this time. On the drive back to the town, he told me that the visit was profoundly eye-opening and enlightening for him. Until then the workers were just some figures on his financial sheets, but now they were real people to him. Even today, three years after our first visit to the refinery, he has kept his promise and takes time every three months to visit one of the factories of the organisation and carries out a safety visit.

When you are on a safety visit, you are mainly on the look-out for unsafe acts, and not only for unsafe conditions, because it is now clear that unsafe acts cause most of the accidents. Think back to the last time you read the report of a safety visit (must have been called safety audit) of someone to a plant, or if you have not read any until now, let the safety department send you some. You will see, as I have done many times, that the items noted and discussed are only unsafe conditions. They are about pumps not having guards, or electrical cables lying around, or ladders with a damaged rung, etc. I can guarantee you that you will hardly find a single entry about human behaviour. People are not accustomed and trained to watch people. For example, a group carrying out a safety audit will notice a dripping pipe wetting the floor, but not notice just next to it a person balancing himself precariously on a ladder and trying to reach sideways a valve that is nearly out of his reach. Or they will notice a disorderly heap of empty cartons on the floor, but not the man who is opening them with a knife, which he is pulling towards his own body. And even if one member of the group carrying out the visit does, he does not have the courage to talk to the worker.

I once gave a presentation to the managing board of an international instruments manufacturing company. I talked about safety visits, and asked if they had ever gone on a safety visit. Being an international safety-conscious company, all 12 board members murmured that they already had done something like that. Then I asked whether they had talked to a worker they had seen doing an unsafe act. Their answers were inaudible. So I was cheeky enough to suggest, "You were scared of talking to them, weren't you?" Nobody replied to that. Most of them just looked down at the notes on the table in front of them, and some twiddled their thumbs.

I defused the embarrassing situation by stating that they were not the only ones who had this problem. Nearly all managers suffered under this reluctance. The reason being that one is never sure how the worker would react. Workers were rough people and they may react impolitely or aggressively. Or they may ask whether you knew how to do it better. They may even brush your remark aside with a wave of the hand, saying that they had done it this way hundreds of times, and they had had no accident, so what's so wrong about it? If you, my reader, too had an experience like this, i.e. reluctance to talk to a worker doing something wrong, then rest assured that it is entirely normal, and that you are not alone.

This is why we have to learn how to talk to workers. There is a certain method of doing this that we shall be discussing later. It is easy to learn and will smooth the way to better communication between yourself and your people. There is no trickery in it, just developing a change of attitude in you and in your workers.

Unsafe acts, actually, are very simple acts. They are acts like not wearing safety equipment (gloves, glasses, shoes, ear protectors), climbing on a ladder that is shaky, smoking where smoking is not permitted, putting hands in rotating machines, speeding with fork lift trucks, cluttering up the workplace, not wearing safety belts, etc. These are things anybody can notice and recognise as unsafe acts. One really does not need to be an engineer to discover such things. But because seeing and noticing are two different things, one needs training to do it properly.

Seeing is a physical activity. Light rays fall through the eye's lens on the retina, and the image is transmitted to the brain. Physically, you see everything that meets the eye. But you do not notice everything, because noticing is a mental activity. Your mind acts as a filter, and you notice only things in which you are interested, or that stand out in the mass of visual information received by your eyes. You would know many details of things you have noticed, but none of the things you have seen, and not found worth your attention.

It is similar to hearing and listening. Your ear receives all the sound waves emitted around you, but you manage to listen only to things that are important to you. You can stand in a crowd at a cocktail party where everybody is talking, and manage to listen only to the person you are conversing with. However, if that person is talking of things totally uninteresting to you, you do hear him, but your mind starts to wander, and you stop listening.

An elderly couple was sitting on a park bench on a sunny afternoon, enjoying the weather. The woman was knitting with her head down in her work, a small sweater for her grandchild, and the man was just watching the world around him. Suddenly a pretty woman walked past them. The man looked at her appreciatively; the woman continued knitting with her eyes on the sweater in her lap.

He: That's a good-looking woman there!
She: Which one?
He: The one that just went past.
She: Many women went past here. Which one do you mean?
He: I mean the good-looking tall one that went past.
She: Blonde or brunette?
He: I think she was a brunette.
She: Many brunettes went past here. Which one do you really mean?
He: Uh ... well, I mean the one that *just* went past.
She: You mean the one with brown shoes, black stockings, a nonmatching green skirt, a crumpled pink blouse and dishevelled hair?
He: Well ... I dunno, just the woman who went past here. She had a pretty face.
She: No, I didn't see her that well.

I guess that clarifies the difference between seeing and noticing. Noticing also means registering the fact in your mind. Noticing is easier if one has a sort of check-list of what to look out for. We shall look at a list like that in a following chapter.

13
Safety Visits – The System

Safety visits have the following basic requirements. They should be:

- regular,
- preannounced,
- punctual and,
- without fail.

Let us discuss each of these points.

Regular means not capricious or casual, but carried out at fixed intervals in time. This could be once or twice a month, or even once a

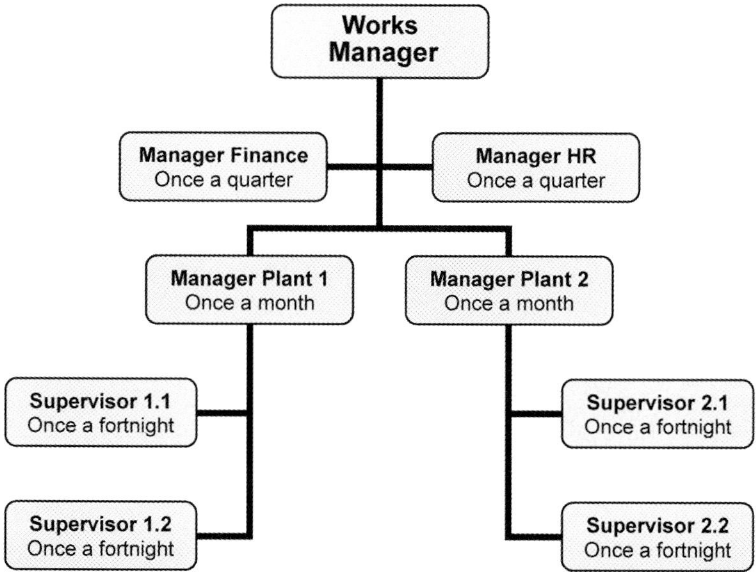

Figure 11 Frequency of safety visits

Managing Safety. Kishor Bhagwati

ISBN 3-527-31583-7

quarter. Let us take as an example a factory, led by the works manager and his staff, and with two plants each having a plant manager. Both plants have two supervisors each.

For a simplified organisation of this type we would have 9 persons carrying out safety visits. The frequencies with which they carry out the visits would depend on their managerial level. I recommend the following frequencies for the organisation above:

Managerial Level	Frequency of Safety Visits
CEO, Board Members, and Managers HR and Finance	at least once every quarter
Works Managers	at least once a month
Plant Managers	at least once a month
Supervisors	at least once a fortnight

To save you the trouble of calculating, this turns out to be 140 visits a year, i.e. about 10 to 12 a month. But not for the same section of the plant! The number of visits are spread over a larger area. For example, the works manager, who visits the whole site, may be visiting a particular section at the most once in six months. Apart from the works manager and the staff functions, production people visit only their own areas of management.

The term "at least" in the table above means that this is the minimum frequency of visits. Depending on the amount of interest and involvement you have, you may want to invest more time, and double the frequency. In companies with excellent safety records this is usually the case. In larger corporations, the head office staff, including the CEO, also takes part in regular safety visits at their different factories.

If you compare the number of visits with old-style inspections, you may feel that the number is too high. But we are not talking about such inspections. The safety visits are short, and limited to an area one can manage in about twenty minutes to half an hour. As a matter of fact, a supervisor, who does safety visiting most frequently, may not even need more than 15 to 20 minutes. He goes through the plant often anyway. The works manager, when he comes, would and should take more time, because he is there so seldom. His is more a hand-shaking, care-taking and shoulder-clapping tour rather than an investigative visit. His personal area of safety management lies at a higher level, namely with his managers who report to him.

With this number of safety visits one needs someone to coordinate them. Otherwise there could be a clash of two managers carrying out their visits at the same time in the same department. This is where the safety department comes in. The planning of safety visits is the job of the safety department. It first makes a master plan, distributes it to the people concerned and makes necessary changes if a visit time has to be changed. And then it distributes the plan to the individual departments about three months in advance.

With that we come to the second point: *pre-announcement* of visits. This may sound very unusual to those who have already carried out an inspection. Wouldn't the people then work extra carefully when they know that you are coming? Wouldn't they clean up before you come, so that the shop-floor looks good? Yes, they would! And why not? What is so wrong about it? As you have seen, the frequency of the visits is so high that there is no use carrying out a crash programme that can be forgotten once the visit is over, because the next one could be on the day after. Your main purpose in preannouncing your visit is to show to the people that you are not coming there secretly to catch them with "the dagger in their hand", but are completely open with them. You are not there to catch culprits and scold those who are doing wrong things. You are just there to discuss with people their safety, and how they can contribute towards its improvement. Above all, you are there to show them that you care! That is what the visibility of our commitment to safety means. Sneaking onto someone to find fault with him is exactly the wrong thing you could do if you want him to believe that you care for his well-being.

Let us now talk about *punctuality*. If you proclaim that safety is important to you, then "appointments" made for safety visits should be as important to you as all other business appointments you make. We decided to put safety at the same level of importance and priority as we put all other business matters like production, sales, profits, etc. Apart from this, whereas your other appointments concern a few people who are your equals or superiors, your safety-visit appointments are watched by a larger number of people who consider you as their superior. The former will not be disappointed if you do not keep your appointment at the given time, only irritated or angered at the most. You would always take care to see to it that your inability to keep an appointment is communicated to others in advance, and a new appointment is fixed.

People on the shop-floor think differently. Your remaining away from the scheduled visit would definitely disappoint them, because they would feel that the safety visit is not important enough for you, and they will feel neglected. That is why punctuality is very important to propagate the message of your earnestness in safety matters. Remember the example of the works manager who asked that the appointment with an important customer be shifted, rather than reschedule his safety visit? Sometimes it could happen that there is something so important that you just cannot carry out the visit at the scheduled time. In these cases, just as with a business appointment, let the people know of it in advance, and reschedule the visit for one of the next days. Never ever let it just be called off. Give a safety visit the same priority as other appointments, and do not let it be cancelled at the drop of a hat, as I have seen many managers do.

This is the meaning of the fourth point, "*without fail*". If you have to carry out 12 visits a year, do not carry out only 11 or 10. Carry out all 12. Don't forget that your name is on the notice board together with the dates and times you are expected to make the visits. What message will you be sending to your people if you carry out only half of the scheduled visits because you had some "more urgent" things to do, or, between us, you just did not feel like going on a safety visit? The message will be: Safety is not that important to me, or, safety is less important to me than many other things I have to do, and I always give preference to other matters. Remember: Safety is only as important to people as their superiors make it. If you do not place safety on a high level, do not expect your people to do it.

Once you learn how to do safety visits properly, it stops becoming a chore, and starts becoming a nice experience. With every visit you make more and more friends, improve the working morale at site, and see smiling faces greeting you every time you come. You get closer to your workers, gain their respect and soon have a workforce that will be prepared to go through fire for you. If you give workers the opportunity to vent their grievances in a one-to-one conversation with you, you do not force them to seek the unions to ask them for help in solving their problems. This results in an amiable atmosphere between union or works council representatives and the management, each helping the other to progress the company towards better productivity and profitability.

When you go on a safety visit, forget all the things that are pending in your office, look forward to seeing friendly faces greeting you, and look forward to the feeling that you are helping people who are dependent on you and expect you to take care of them. Leave your mobile phone in the office, and tell your secretary not to let anybody disturb you on your rounds. For the next hour, safety is uppermost in your mind, the only thing you want to concentrate on.

When you are out in the plant, there will be many who would like to take the opportunity of buttonholing you and discussing production or other problems with you. The superintendent of the area you are visiting would like your decision on an issue that is burning in his mind. At such moments, tell him firmly that you do understand that his problem is important, and that you will discuss it with him later, but for the present you only have safety on your mind, and you would not like to discuss anything else but safety.

Safety will only be as important as *you* make it.

14
Safety Visits – The Procedure

Going on safety visits is very easy. As compared to conducting meetings, where you have to prepare yourself in advance, or giving presentations where you have to spend time working on your PowerPoint slides, making a safety visit does not need any special preparation. After all, you are mostly going to do some listening, and very little talking. And even the talking you will be doing does not need any preparation. It is spontaneous; it just puts into words what you think. Once you have learned to do safety visits in the sense they are meant to be done by following the instructions in this book, you do not have to worry about any worker backtalk or odd replies either. So enjoy your visits – they really are very satisfying.

There are certain basic procedural rules one has to understand and accept. These are:

1. Acquaint yourself with the site's safety rules and follow them strictly.
2. Build a small team of only 2 to 3 persons for the visit.
3. Enter the area to be visited after waiting a few seconds on the threshold.
4. Ask the worker whether he has time for a few minutes.
5. Stick to the talking sequence described in the next chapter.
6. Note down any commitments you make.
7. Never part from a worker without thanking him for his time.
8. Write a safety-visit report.

1. Follow all safety rules

When going on a safety visit, you have to follow all the safety rules of your own site. Thus if the safety rule says "No smoking" in the

Managing Safety. Kishor Bhagwati

ISBN 3-527-31583-7

plants then, if you are a smoker, leave your cigarettes and lighter in your office. If the walkthrough requires safety shoes, safety glasses and safety helmet, wear them. See to it that you are issued your own personal protective equipment, and have it in your office cupboard all the time. You have to be protected when you go through a plant, because accidents do not care whether you are a worker or a manager. They hit both equally. Apart from that, on your safety visits you have to set an example and be a role model. However inconvenient you might think safety shoes are (the newer ones are not that inconvenient), wear them. Just think that it is your own foot you are protecting. You cannot expect your people to work safely by wearing protective equipment, if you yourself do not wear it.

At a chemical company I was consulting, the safety manager moaned that the only person walking around the site without a helmet where wearing of safety helmets was compulsory, was the works manager. He had requested him very often, but the works manager just thought safety rules did not apply to him, because he was not doing any work in the plants. On my asking whether that gentleman had a wooden head that was unbreakable and impenetrable, I got another big sigh.

Three weeks later was the annual meeting of all the employees of the site. This was a regular feature in this country. As per my recommendation, the safety manager prepared something for the works manager. At the big meeting, where every employee was present, the safety manager went on the stage after the works manager had finished his talk, and presented him a nice new helmet with two golden stripes around the circumference and the manager's name and title painted on it in red. He offered it to him with the words: "This, sir, is your personal helmet for your next round in the factory." And it worked. Never again was the works manager seen in the plants without *his* helmet on his head.

Trouble arises when you have to take visitors around the plant. The rule of wearing the prescribed safety equipment is equally valid for visitors as for own people. The normal argument goes: OK, one can furnish everyone with a visitor's helmet that has an adjustable headband to fit all sizes, and also plastic safety glasses that fit all. But what do we do about shoes? One can't have a storehouse of safety shoes in all various sizes for all visitors. It is also a question of hygiene. And one can't just discard such expensive shoes after they have been worn once.

The objection is quite justified, but there are also solutions to this problem.

The first solution, for example, is designating and marking certain safe walkways that the visitors are not supposed to trespass. As long as one remains within these markings, one need not wear safety shoes. Such a solution is possible in smaller plants in buildings, but difficult to realise in large open plants. The second solution, therefore, is to have a few shoes in all the common sizes, and issue cheap white socks to wear when putting on the safety shoes. The socks can then be discarded, and the insides of the shoes sterilised with a spray. The third solution for plant visits on “open days”, where the families can visit their relative’s place of work, is to stop the plant completely during the visiting hours.

2. Build a Safety-visit Team

You can always carry out a safety visit alone, but it is advantageous to have one or two persons from the plant with you, in case any explanation is required. Arrange to meet these persons with whom you would be carrying out the visit in front of the plant entrance. Who should be the persons accompanying you? First, the person responsible for the area you are going to visit on that day, and second, a worker from that area. Thus if you are going to visit a maintenance shop, the persons accompanying you would be the workshop supervisor and a fitter, electrician or a welder. Nobody else! No one from the safety department or the works council. Certain countries’ laws require a team for such safety rounds that includes the plant safety person, the site safety person, the works council person, the medical person, the plant manager, the site security person, the environmental person, etc. You then have a wave of persons going through the plant, finding faults here and there, discussing amongst themselves whether a particular condition is safe or unsafe, and noting down things openly in front of the workers for their later meeting. That none of them ever looks at a worker doing something unsafe is quite obvious. That is not what the law demands of them. Fortunately, these Spanish Inquisition types of safety rounds feared by the workers are to be carried out only once or twice a year.

The worker accompanying you should be a different one at each safety visit of the area. All the workers of a department have to be on the team in a continuous cycle. That the workshop supervisor should be with you is understandable. But why the worker? The reason is very simple. Today the worker who is with you plays the part of the observer. He sees things with a different eye, with the eyes of a visitor. He notices unsafe acts immediately, because he himself may have been doing them. He becomes aware of the short-cuts one takes to avoid effort. From his present role as an observer for unsafe acts, some of the things that are happening are eye-openers for him. And tomorrow, when he is again doing his normal work, he suddenly becomes conscious of an unsafe and risky action that he intends to carry out, and his mind says: Stop. As an observer, he would have found a colleague doing the work in this manner objectionable. So why do it now himself? This has a very sobering effect on him, and he stops doing risky things voluntarily. A great leap forward for safety!

The safety visit starts with meeting the team outside the plant, and is followed by a round of a limited part of the plant. Do it in smaller pieces, rather than taking on the whole plant at a time so that you do not have to rush through the plant and cannot take time to talk and listen to workers. Take your time when carrying out a safety visit. The time of your visit and the section you are going to visit have been pre-announced, and people are expecting you. Do not disappoint them by carrying out a superficial "flying visit". Your safety visit should not degenerate into a sight-seeing tour of the Europe-in-Seven-Days-type ("Darling, what day of the week is today?" "Tuesday." "Oh, then this must be Rome!").

3. Entering the area to be visited

Before you enter the area and start your safety visit, wait at the threshold for a few seconds, and let your eyes roam. Once you are seen by the people, the world will change immediately. A worker will push down his face-shield, another will quickly wear his glasses, a third will put on his gloves, a fourth will push his open drawer shut with his thigh, a fifth will walk away from a rotating machine, etc. The purpose of your waiting a few seconds is not to "catch" anybody, but to find out with whom you would like to talk. Do not let your eyes roam like that of an inspector with a grim face, trying to see whom to punish, but just look quickly and blankly at the area in general. A manager told me that he looked at the area as if he was looking at beautiful mountain scenery, with a happy expression on his face. You can find your own way, but please try to avoid all signs of authority.

4. Approaching a worker

Before you start a conversation with a worker, you have to approach him. With your sweeping glance from the threshold of the section you may have selected the person with whom you may want to talk. Or, if you did not discover anyone in particular to whom you would like to talk to, just move around and find the next person carrying out an activity. Stand away from him at a respectful distance, so as not to disturb him in his work, and start talking with him when you think he can interrupt his work. During the time you are waiting, do not stare at him, but look around at others; otherwise you may make him nervous and liable to a false move that could result in an accident.[26)] When the proper time has come, go nearer to him, greet him and ask him whether he has some minutes to spare. If you are very high up in the organisation, and he may not know you, then introduce yourself.

A top manager in my course introduced himself to a worker with the words, "Hello, I am Dr Suzerain, the General Manager of this company and your topmost boss. I would like to talk to you about safety." The worker had seen his "topmost boss" only once, but never so near, and he had never talked to him. He started asking himself: God knows what he wants now? He got very nervous, and the conversation did not proceed well at all. To all questions of Dr Suzerain, he just answered in monosyllables. Upon being asked whether he had any problems, he just nodded his head. He did not want to get his supervisor into trouble, and so he just clammed up. The right way for Dr Suzerain would have been: "Good morning, I am Suzerain and also work for this company. Do you have a few minutes to spare to talk to me about

26) A friend of mine told me that he was in reality a very careful and good driver. But whenever his wife sat next to him, not saying anything but just staring with a worried look through the windshield or in the make-up mirror, and often throwing a sideward glance at him, he invariably drove over some obstacle, or passed a red light. There must be a scientific name for this phenomenon, but I do not know it.

safety?" This followed by further sentences we shall be discussing in the next chapter would have resulted in what we want out of a safety visit.

5. The talking sequence

Although we shall be going into the details of this point in the next chapter, I want to stress here that this is the most important part of the safety visit. You can be perfect in all the other points, but if you do not practice this correctly, you will not have the desired success.

6. Noting down

Never have a note-pad or a clip board in your hand when making your safety-visit rounds. Remember that we want to eliminate everything that may awaken the impression that this is a fault-finding mission or inspection. No meter-maid image, please (however lovely Rita maybe!). What you should carry with you is a small note-book in your pocket. This has to be fished out for making a note whenever you promise someone that you will look into the matter you were discussing. This way you will not forget what you have promised, and the worker will see that your words are not just empty sounds, but that you mean what you say. The image you project will make or break the trust you will gain from your workers.

Once you are back in your office, use another note-book to note down what you have seen, with whom you have talked and about what. This is your secret booklet. These notes serve as an *aide-mémoire* for your next round. The purpose is that you would like to have – and give – feedback to the points you had discusses last time. Above all, it is useful to know who deserves special praise from you. You can light a flame of willingness, but the thing to keep it burning is the oil of praise. The praise should, of course, be honest. False praise and empty words will be immediately seen through by the workers. A manager who goes around, clapping the shoulders of his workers and telling each of them that they are "doing a great job", is as unbelievable as the Egyptian bazaar dealer who tries to sell you two skulls of Cleopatra, a smaller one when she was a girl, and a larger one when she had grown up.

Honest praise and recognition is the fuel of motivation.

Remember the time when you were in primary school and the teacher had asked that for the coming Easter each should make a painting of an Easter egg? You sat at home with your colour crayons or felt-tip pens and made a beautiful multicoloured egg with stripes and dots on it, and took it full of pride next day to the school. And just as you wanted to show it to your teacher, she said she does not have time to look at it, and never found time to look at it later either. How disappointed did you feel? You had so much hoped that she would appreciate what you had done, and even praise you for the choice of colours or something like that. But not even looking at it? You are never going to do anything more for her in future. As it is, she doesn't care whether you do it, or not!

Let us transfer this situation to your safety visit. During this safety visit you had seen Jack grinding a work-piece without safety glasses. So you talked to him, convinced him, and he promised to always wear his safety glasses whenever he did any grinding. On your next safety visit, he is grinding again, and has his safety glasses on. You go past him, and do not say anything. Either because it is not your "style", or because you just forgot. What is his reaction? He knows now that you just don't care whether he has his safety glasses on, or not. So why wear them next time? You wouldn't notice it anyway.

On the other hand, if you had looked up your *aide-mémoire* booklet before leaving your office, noted in your mind that you had talked with Jack the last time about wearing safety glasses, and made it a point of looking up Jack again and seeing whether he was now wearing his safety glasses or not, and made a positive remark, i.e. a bit of praise if he was wearing them, it would have generated in him a feeling of recognition of his conversion to safety, and he would continue working safely after that. One does not have to show one's appreciation the way we do where I live in the French part of Switzerland by kissing him thrice on the cheeks; you can just give him the thumbs-up sign and say: "Thanks, Jack, for using glasses. I appreciate it." Show him that *you have seen* his positive intention. That will be enough.

7. Thanking the worker

Inspectors always feel they are not imposing upon anybody's time. What they are doing is so important, that all other things have to take a second place. Or have you met a policeman who caught you for speeding ever thank you for the time you spent explaining to him how you would miss your plane if you did not reach the airport quickly, and finally even paid the fine on the spot, all this costing you at least ten minutes? Inspectors, who are symbols of authority, do not have to do this. They have to keep a straight face to show that they do not have any positive feelings towards you that may lead them to lower the fine, or even let you go. A smiling inspector is as much a contradiction in term as a smiling jogger.

And that is exactly what you do not want to be, and also do not want to be seen as. As we shall see, we have to have a much higher respect for our workers than a traffic cop has for us. And that is why you should thank him for the time he has taken to talk with you. This may be just a formality, but it fosters your relationship, and, above all, shows him that you respect him and are not an inspector in the normal sense of the word.

8. Report writing

At the end of the visit, the team should sit down in the cabin of the area responsible person and note down the points it has noticed. The actual typing of the report can be done by the department later. This will not be like a newspaper report; hence lengthy descriptions are not necessary. The report is meant to be a document solely for use by the department and for the purposes of follow-up. One-liners will be sufficient for this, because all concerned know what one is talking about. A one-page safety-visit report form is given on page 109. More than a page is never necessary.

The report is divided into three parts:

1. Background information: date of visit, time of visit, plant or section visited, names of the visiting team members, number of persons employed in this section, number of persons approached and talked with.

2. Description of things noticed: what you and your team saw, whether it was an unsafe condition, or an unsafe act, what you said to someone who was performing an unsafe act, what were the praiseworthy items you found, etc. NOTE: The names of the persons you saw or talked to are not to be mentioned in the report.
3. Follow-up: You start here with first noting down the positive things you have seen, e.g. cleanliness. This is followed by what action needs to be taken to defuse an unsafe condition or to improve an unsafe act, which of the team members takes over the responsibility of managing the completion of the action decided, and by when? The deadline is to be given by the person who has taken over the responsibility of completing the task.
4. No person who is not present is to be given that responsibility, and it is never to be given to a department. The name of a person is essential for follow-up.

All action points must be accompanied by the name of the person accepting the responsibility for its completion, and the deadline he himself has given.

5. The report is then displayed on the notice board in that particular section, and a copy is sent to the superiors and to the safety department.

The purpose of displaying the report on the notice boards is, first of all, to let others know the types of unsafe acts one has found in that section that they should avoid in the future, and also to let everybody know who has accepted the job of remedying a particular situation and by when. This way, the manager does not have to continuously follow-up on the items to be finished, but peer pressure enforces the deadline. For example if Bill was supposed to get the rung of a ladder repaired by the 25th of March, his colleagues will remind him that the date was approaching, and that he should hurry up with it.

The report is distributed to the departmental manager, the plant safety person, and to the site safety department which keeps a record of the visits. A "need to know" philosophy is to be observed. The site safety department collects the data on the fulfilment of safety visits by managers, and the safety professional presents a table with the names of the managers and their scheduled date of visits, as well as their date of completion, once a quarter at the management meeting. If a manager has neglected his duty of visiting for a few months, he has to justify it at this meeting. "No time" is an unacceptable excuse.

A word about deadlines. A deadline should always be a date, and not a week or a month or a quarter. Thus a deadline can be "15 March", but not "mid-March" or "week 11" or "end quarter one" or "in the spring". Such hazy and vague deadlines cannot be followed-up for checking the completion of a job.

In Appendix 3 you will find two completed safety visit reports. Decide for yourself which report is better and in line with what has been said until now. That of Mr Saras, or that of Mr Malfoy? You will find hints for quickly assessing the quality of safety visit reports on the first page of the appendix.

<table>
<tr><td>Safety Visit Report</td><td>Plant:</td><td colspan="2">Section:</td></tr>
<tr><td></td><td>↓ Name</td><td colspan="2">↓ Position</td></tr>
<tr><td>Safety visit by</td><td></td><td colspan="2"></td></tr>
<tr><td>Accompanied by</td><td>1.
2.
3.</td><td colspan="2"></td></tr>
<tr><td>Visit date</td><td></td><td colspan="2">Duration:</td></tr>
<tr><td colspan="2">No. of persons talked with</td><td colspan="2"></td></tr>
<tr><td colspan="2">No. of persons working in the visited area</td><td colspan="2"></td></tr>
<tr><td colspan="2">Findings S = Safe U = Unsafe</td><td>Act</td><td>Condition</td></tr>
<tr><td colspan="2">1
2
3
4
5
6
7
8</td><td></td><td></td></tr>
<tr><td colspan="2">Totals</td><td></td><td></td></tr>
<tr><td colspan="2">Acts observed / Conditions observed =</td><td colspan="2"></td></tr>
<tr><td colspan="2">Follow-Up of Items listed above</td><td>Person</td><td>Deadline</td></tr>
<tr><td colspan="2"></td><td></td><td></td></tr>
<tr><td colspan="4">Distribution: Plant Manager (*), Plant Safety Person, Central Safety Department, Plant Notice Board</td></tr>
</table>

(*) This form is for supervisors. For managers, use the next higher position in line management

15
The Art of Talking

Before we discuss the art of talking, let us clarify a few things that would help you use the techniques described below properly.

The worker you are going to talk to you is a person selected by your company. The company had a choice when choosing him, and took the best of the available candidates. Your HR department did not just open the gates and let anybody and everybody come in and pushed a contract into his hands. He was selected on the basis of certain criteria that fitted your company's requirements. He was confirmed after a probationary period, during which he proved that he could satisfy your expectations. You have a person here who is intelligent, smart and hard-working. So please treat him that way.

The second thing you have to realise is that he knows his job and what he does much better than you do. The intelligent person you have selected does not switch off his brain when working. He knows all the apparent and visible hazards of his job, and those he cannot know, the company's safety training department has taught him, e.g. hazards of handling dangerous chemicals. He is fully aware of the problems that exist, and has thought about them during his work. He has often found that the procedure of working or the machine he works on may cause him an injury, and has even thought of some improvements to prevent it. But very few workers pick up the courage to talk to their superiors about these improvements on the machine or on the process. Your company may have a system of so-called "idea management" through which, if he submits his idea in writing, he could even get some reward! Easy to say for us desk workers, for whom writing is nothing special. Remember, however, that many of the workers have not written much since they left the school, and feel very insecure about spelling and syntax. Rather than expose their ignorance, they take the easier route of not writing. In this way, many good ideas that could have increased safety are lost.

Managing Safety. Kishor Bhagwati

ISBN 3-527-31583-7

Once, in a plastic processing plant near Paris, I was accompanying the works manager on his safety visit. We came past a machine producing tear-off type of plastic bags for grocery stores. You know the type where a hundred or so bags are welded together at the top as a bunch, and have two large holes on the welded part to enable them to be attached to a stand (like in a flip-chart). One pulls off each bag individually by tugging at it. The machine was throwing out the bag-bunches rapidly, and the worker had to insert each bunch on two large vertical rods till they reached the height of a carton. These rods had quite sharp tips. The works manager asked him casually whether he saw any hazard in his work. Yes, he said showing a scar on the inside of his lower left arm, he got it through one of these sharp tips. He was worried about those tips and was afraid that one day one of them was going to penetrate much deeper into his arm or wrist and hurt him very badly. The works manager asked him what solution he suggested to remove the hazard. He asked in return, why did the tips have to be so sharp? After all, the holes were nearly 3 cm in diameter, and the rods could be blunt. He was quite right. The works manager ordered the tips to be changed immediately. And then he turned to the worker and asked, why hadn't he told it to anybody? His answer was: "*Personne ne m'a jamais demandé!*" (Nobody had ever asked me!)

Safety visits are for asking such questions. Workers may not be confident in writing, but they all can talk. And this is what we have to exploit on safety visits. Remember that we are out there to listen to people, and not to tell them much. We want them to tell us where the shoe pinches, or how they can make their workplace safer. *They* know best how to get rid of a hazard, not the engineer sitting in his office. And their solutions are the cheaper ones, because they think in simpler terms. They do not think in terms of electronic gadgetry or computer-controlled systems, as an engineer very probably would.

In a cosmetic factory I was making the rounds with the safety officer, and we were walking through the plant that made shampoos. One equipment was a 5000-litre agitated vessel with pipes connected to it and having a manhole on the top of about 60 cm diameter. The main ingredient of a shampoo is a surfactant, similar but milder to the ones used in dish-cleaning detergents. To that are added all those magical additives, colour and perfume. The major component, the surfactant, was fed into the vessel from a storage tank via a pipe connected to it. All other ingredients had to be put into the vessel manually through the manhole. As all these ingredients and their amounts were supposed to be secret, a group of people in the formulating department put each of those ingredients in a pail, and the pallet with the pails was delivered to the plant. The worker was to add the secret ingredients one after the other in a given sequence by lifting the proper pail and emptying it into the vessel through the manhole. As all the ingredients, being meant for a shampoo, were not hazardous, no particular care was necessary when handling them.

I asked the operator working on the vessel, whether he saw any hazard in his work. He said yes, there was one point that worried him often. When lifting the pail of one of the main additives, which was present in all shampoos and was a slippery liquid, it often slopped over and made a puddle on the floor. He had often slipped on the slippery floor, twice hitting his head against the railing behind him. Those two times he got away with a minor concussion, but he knew that one day he would not be able to stop his fall enough, and would hit his head heavily against the railing or the floor, hurting himself badly. The reason was that this slippery ingredient, about 20 kg of it, was in a pail that was filled nearly to the brim, and every time he lifted it, some material fell out.

The safety officer, who was with me, started explaining that they were thinking of substituting the open pail with a special stainless steel pail with a lid and a snout .... I stopped him and turned to the operator and asked him whether he can suggest a solution. His answer was amazingly simple. He suggested using two pails instead of one for the material, so that both pails were half full, and there was no danger of anything spilling out! Brilliant, isn't it? I have experienced such things quite a number of times, and it would be wise on the part of the managers to tap this innovative resource, rather than delegating others, who do not carry out this job, to find a solution for the problem.

Use this approach when you want to find a solution that is simple and inexpensive. Firstly, it has the advantage of instilling pride in the worker for having found a solution to a problem himself, and secondly, it is a solution he will not object against, because it is his own. You do not have to convince him of using a solution that he does not think is appropriate. The idea is already sold in advance. No selling needs to be done by you.

Your conversation with a worker should follow the outline given below. You are going to talk to him either because you have seen him doing something in a safe manner and you want to show your appreciation for his attitude by praising him, or you have discovered that he was carrying out an act that could cause him injury sooner or later. The first case is easy. It is no problem telling somebody positive things. Showing a person appreciation for his safe method of working is as important as talking to a person who is working unsafely. Because it is so easy, one often forgets it. One takes it for granted that a person working properly does not need a pat on the back. That is where we are wrong. Praise is recognition that works as an accelerator for further progress. A flower in bloom needs water as much as a flower that is hanging its head. Do not, therefore, forget to give small encouraging remarks to such persons when you see them.

Let us now take the case of a worker who is carrying out an *unsafe* act:

1. You ask him what safety precautions he has taken for carrying out the job, or in other words, what has he done to see to it that he does not injure himself. If the answer is very general, ask him to be more detailed. For example, if he says he has taken all the safety precautions, ask him which ones. You can prod him by mentioning that you had noticed that he was wearing leather gloves and/or safety shoes. Let him list all the safety items he is wearing or using. The purpose of this question is to make him aware of the fact that he is already doing something for his safety.
2. You show your appreciation for his safety awareness.
3. Next you ask him why he takes these precautions. His answer would be: Not to hurt himself. You confirm the answer by agreeing with him, and mentioning that it is a wise thing to do.
4. Then you ask him whether he is aware of any unsafe act he has been doing. If he is honest, he would agree that there was something he was doing that should have been done in a safer manner. If he feels he has not done anything unsafe, then you can point out that you, however, had noticed that he had, for example, not worn his gloves when handling the very sharp knives of the cutting machine. Remember, you are just speaking about your observation, not reprimanding him. What you have discovered is what is known as The Force of Habit. People do many things unconsciously out of habit,

never thinking about them. It is not ill will; it is just unawareness of the moment. Be, therefore, considerate when talking of this gap in his safety behaviour. What we are striving at is a change in this habit; from a habit that could hurt him to a habit that would spare him injury.

5. After he has agreed that he did do something unsafe, time has come to discuss with him the consequences of his act. Never forget that you are basically here for listening, and not telling. First ask him about the consequences of his unsafe act. Insist that he takes the worst-case scenario. If he does not agree to this, ask him whether he can guarantee that the worst case will never happen to him? Does he not have a car accident policy or a sickness insurance policy? Would he be able to guarantee that he would never need them? All this should be done in a polite enquiring tone, and not with a raised finger.

   Let him describe this worst case. Do not let him argue that it has never happened until now. That is no guarantee for it not happening the next minute! The fact that it could happen is enough. In the case of knives, it could very well be that he loses a finger or two (if a knife starts slipping and he holds it hastily). Things like slips and falls could even result in deaths! You do not have to give such drastic examples if not needed, but you should prompt him to think further, remembering that all the thinking should be done by him.
6. The question to ask now is: would he like to suffer the injury he just described? How would he feel if he, say, had to walk around the rest of his life with two fingers less? He may not be able to do so many things he would have liked to. His answer would definitely be "No".
7. And now the most important question to him: How can he prevent such an accident happening to himself? What does *he* suggest? No answer or solution from you! *He* has to think and tell you what he can do, or what can be done. Remember the preamble to this chapter. The best answers are there where the work is being done, and where the hazard exists. And the best solutions are of those who face the hazard every day, and who have the most to lose if this hazard gets out of hand. The added advantage, as mentioned, being that the idea is already "sold" to him before you even hear of it! In the case of the sharp knives, he may suggest that he could wear tough safety gloves.

8. You agree with him and commend his idea. Safety gloves would surely prevent him losing a finger or two. And now you want something from him. You want from him the assurance (use the word "promise", if your culture allows it) that he would always wear safety gloves when handling the knives. Not for anybody else's sake, but for his own. Explain to him that in case of an accident, he and his family would be the only losers. He would be the one who draws the shorter match, and nobody else. He and he alone will have to bear the burden of pain, suffering and maiming. Hence whatever he does in future, he should do it for himself and in his own interest.
9. Shake hands upon parting and leave him with the feeling that you have a personal interest in his well-being and that you consider freedom of injury for your people as your personal responsibility.

If you, however, see a person actually doing something that could very plainly hurt him and you want him to change it immediately, then you can leave out the first three steps, and plunge right into the discussion of the unsafe act. Let us take an example, and do it this time like a stage script:
You are walking through the plant and see a worker using a flex grinder on metal. Sparks are flying all over the place, but he is not wearing his safety glasses.

Let us first do it "*the good old way*":
You tap him on his shoulder or talk to him standing very near to him and say:
You: Jack, I see that you are breaking the rule again. You are supposed to wear safety glasses when you work with a flex. Why aren't you doing it? If I catch you working this way again, I shall see to it that you get a written warning.

And now the "*new*" way:

You stop a few metres away from him, try to catch his eye, and ask:

You: Jack, could you please stop what you are doing and come over here for a minute?

Jack switches off the flex, puts it aside and comes over to you. The purpose of your saying this is to get him out of the zone of danger.

You: Jack, I saw you grinding just now and lots of sparks were flying around.

Jack: Yeah, they always do.

You: I am sure you know that these sparks are red hot metal grains.

Jack: Yes, they did tell us something like that during flex training.

You: Now tell me, what would happen if one of those red hot pieces got into your eyes?

Jack: That wouldn't be nice, would it? It would damage my eye.

You: You are right. I know a person who lost his eyesight doing work like this. Would you also like to suffer that fate?

Jack: Oh no. Not me. I don't want to lose any eye.

You: How can you prevent it, Jack?

Jack: I guess by wearing safety glasses when flexing.

You: That's a good idea. Why don't you do it right now and then always?

Jack: You're right. I will. *Puts on his safety glasses.*

You: Very good Jack. Thanks. Can I count on you always wearing the safety glasses when you flex in future?

Jack: You sure can. I don't want to lose an eye.

You: I hope that you don't too. Bye, Jack. I wish you a good day.

They shake hands and part.

What was the difference between the two methods? The first threatened, and the second convinced. Which of the methods, do you think, would be more effective, and above all, longer lasting?

Let us now take an example where the unsafe act does not pose an *immediate* danger, but will cause him damage on a long term. You are taking your round in an area that is very loud. Ear defenders or at least wearing of ear plugs is the safety requirement. You see a worker working in this area, but wearing neither his ear plugs, nor his ear protectors. First, the good *old* way:

You: Hey Bill, didn't you see the sign when you entered this area? The one saying that everybody entering this area must wear ear protection? Why aren't you wearing any? I don't want to catch you again without ear protection, OK?

And now, the new way we want to do it in future:

You: Bill, could I have a word with you? Let us get out of this loud area where we can talk.
Bill and you leave the loud area or the building.

You: Bill, I am worried about you. I saw you working there without anything to protect your ears. I am sure you have heard the safety lecture on effects of loud noise?

Bill: Yes, I did.

You: Did the trainer tell you that the loss of hearing through a long-time exposure to loud noise is absolutely irreversible, and that there is no medicine or operation that can give you your hearing back?

Bill: Yes, he said that, but I don't have any problems with the noise here.

You: Maybe your ears are accustomed to the noise, but the noise damages anyway. Only you do not notice it today, but it is slowly and steadily reducing your ears' hearing capacity.

Bill: Why do I need to worry about it today?

You: Bill, just think ahead. One day you will be retired and have plenty of time to go the pub and enjoy a drink or two with your pals. But just imagine, you will not be able to hear half of what they are saying and will not understand the jokes they are laughing about! Or you love music, and you have planned to buy a super hi-fi set when you are pensioned. Forget it. You can save your money and buy a cheap portable radio. You won't notice the difference anyway. And all that just because you are not wearing the ear protection today.

Bill: Is that true? You mean that is what would happen to me if I don't wear these tiny plugs *today*?

You: You are right. But Bill, you decide yourself how you want the quality of your life to be when you are old. Don't wear the protection because of me or the company. It is not I who would get deaf if you don't wear the protection. And the company, believe me, will not stop working because of it. You, and only you and your family are the sufferers. But

> please, I leave the decision in your hands. I am not insisting on you wearing it.

I can guarantee you that Bill will never be seen again in this part of the plant without his ear protectors, and also that he will relate to all his colleagues the conversation you had with him, and recommend them to wear protectors too.

I think the differences in the two approaches are now quite clear. Once you see your visit as an instrument to show your people that you care, and to convince them – by highlighting their own, personal interest – not to take risks, you have done your job well.

We talked of asking the worker for solutions to problems as in the example with two pails. This is worth going through again.

You will be surprised at the simplicity and the economy of solutions to safety problems suggested by your workers. People who earn less money have a different feeling for monetary numbers than do others who deal with large amounts on a regular basis. For the worker, every amount is translated into his personal financial resources. An engineer, planning equipment or a building costing hundreds of thousands or even millions, does not relate these amounts to his personal income. But a person who never deals in these sums which, in his eyes are gigantic sums, always relates them to his income. He, therefore, always thinks of a cheaper solution.

In the 17th century, Northern India had a ruler named Akbar, the Emperor. He had a minister by the name of Birbal, who was known for his intelligence and smartness. Once, however, the Emperor got angry with Birbal, and wanted to fine him. Birbal agreed to get fined, but requested that he be allowed to choose the judges who should levy the fine on him. Akbar agreed after making the condition that none of the judges should be a member of his family or from his circle of friends.

Birbal went on the streets and selected three street sweepers as his judges. Akbar was surprised that a minister wanted to be judged by such lowly people, but he kept his word, and did not say anything. The three judges started discussing amongst themselves, what fine should be levied on Birbal. The first one suggested a hundred rupees, nearly twice their own monthly salary. “What”, said the other, “listen, we just want to hurt him, not ruin him.” The third one agreed that although Birbal deserved a fine for having angered the Emperor, one had to be humane, and suggested a fine of 50 rupees. This went on for some time till they came to a final decision, approached the Emperor, and said, "Ruler of the World, we are aware of the responsibility that is laid on us, and seeing the crime His Excellency has done, we have agreed that he should be fined a sum that would teach him a lesson, and have decided that it should be 30 rupees."

16
Pillar 3: Involving the Worker

Company policies often use the phrase: "Our people are our most valuable assets." If this is true, then you should treat them as such. Assets are used to pay debts and obligations, and our obligation is to the well-being of the worker who works for us. The worker who comes to work for us trusts us, and believes that we will take care of him. Would you like the wife of your worker to worry every morning when he leaves home whether he would return with all his fingers, or return at all, or would you like her to feel confident that he is going to a company or a plant run by you, and that you are someone who sees to it that nobody in your factory gets hurt?

Figure 12 Dilbert cartoon

We can use these "assets" to achieve our goal of superior safety by involving them fully in all decisions concerning safety that we make. I would not stop at safety matters alone. But you can start with safety, and very soon you will realise that this approach has advantages in many other aspects of the running of a factory. You have selected good people, so use their intelligence and, above all, their experience (which you do not have), to find solutions to problems. Why bear the burden

Managing Safety. Kishor Bhagwati

ISBN 3-527-31583-7

all alone on your shoulders, when you have so many able people who will willingly share the burden with you? You must trust your people; otherwise they will not trust you!

As shown in the last chapter, learn to encourage your workers to solve problems themselves. The first requirement for this is that you learn to be humble. The American philosopher Ralph Waldo Emerson once said: "Every person I meet knows something about something more than I do, and in that I can learn from him." If you are talking to a welder, you can be sure that he knows more about welding than you do, and in that one aspect he is superior to you. Similarly, every worker you meet in your factory knows of the work he does more than you do. In face of this, there is no reason to feel superior because of one's position. The angle at which your nose tip is turned upwards and the titles with which you always want to be called by are counterproductive to getting help from people for solving problems.

Except in countries where English is the mother tongue, it is not customary for people to address their superiors with first names. In these countries, a first-name relationship is a special relationship, developed sometimes years after one gets to know each other, and is associated with taking the step from being an acquaintance to a trusted friend. With the first name you change the tense when speaking and writing from the third or second person plural, to second person singular (in German from "Sie" to "Du", in French from "Vous" to "Tu", etc.) One American consultant on intracompany communication invited by my company to lecture in Germany insisted that one calls one's boss by their first name. She was incapable of understanding why this cannot be done.

When talking with your subordinates, avoid using your titles, academic or managerial. It introduces an up-down barrier in the relationship, which is not desirable.

One of the most title-conscious countries in Europe I know is Austria. The wife of a Herr Doktor automatically becomes a Frau Doktor, and it is still customary to spell out all the titles when addressing a person. I was once with a Mr Schmidt, who was the deputy (*German: stellvertretender*) of the CEO (*German: Vorstandsvorsitzender*), and also the Director (*German: Direktor*) of a company division. He was, naturally, also a Herr Doktor ("naturally" because it is more or less a prerequisite for rising in the company hierarchy). We travelled together to one of the plants of the company, where he was greeted by the plant manager with: "Grüss Gott[27], Herr stellvertretender Vorstandsvorsitzender, Direktor Doktor Schmidt." And this not only the first time, but every time he addressed him.

Some days later, Schmidt asked me how to modernise his company, and I unthinkingly suggested getting rid of the titles in the company. The reaction was as if I had suggested reannexing Austria to Germany! Impossible, he exploded; he would feel naked without them. Just by calling him by all his titles puts the people in their place, and he would not like to lower himself one bit and let anyone forget the position and the status he had earned. Needless to say, his visits were not crowned with the type of success they would have been if he had been more humble.

If you ask people, ask them out of a desire to learn about things, and not like a schoolteacher putting an exam question, the answer to which he already knows. You should ask, "Could you please explain to me what you are doing?", and not like the teacher who asks, "What is the capital of Zambia? Huh? Don't you know? Come on, come on."[28] You will be surprised at how willing people are to answer your questions about what they are doing. They will go out of their way to show you how their machines work. Why? Because they are proud of what they are doing, and tickled that a manager wants to know about it and is interested in his work. Questions of this type not only show your interest in what the worker is doing, but also the respect you have for his work and ability. Additionally, it increases your knowledge about what is going on in the factory.

Once they start talking, you may hear things that have nothing to do with safety. Complaining is in human nature. You have to be careful and see to it that your conversation does not start drifting away from safety, and have to bring the train back on the tracks of safety; gently, but surely. There are many ways of doing this, and I am sure you know them. But please do not ever tell him that you are not interested in what he has to say. Just insist that you would like to concentrate on safety for the present.

27) No, he did not call him Gott = God. Grüss Gott is the Austrian and Bavarian way of saying Good Day. It was Charles de Gaulle to whom once his assistant joyously informed: "Mon dieu, we have won!" (By God, we have won!), and he replied, "I am glad about that, but you can still call me mon General."

28) For the "very few" who do not know (and this included myself before I looked it up), it is Lusaka with 1.3 million inhabitants.

Involving workers means letting them participate in decisions that affect them. In any committee or meeting that has to decide on things that directly concern them, see to it that they are represented by at least 50%. The participants should be really from the shop floor, and not just their works council or union representatives. If your factory has more than one unit, and a problem of works-wide importance is to be decided, distribute the membership equally amongst the workers, so that each unit is represented. Take care to observe that it is not always the same person who represents his unit at all meetings. Once you have started making your safety visits, you will find out by yourself whom you would like to be in the group. Asking only the supervisor or the plant manager to select the person is not a very reliable method, as favouritism cannot be prevented. You should know that often there are groups and clans within the plants, and animosity between the clans is not rare.

One of the factories in my company manufactured raw chocolate mass from cocoa beans. The chocolate mass was sold to makers of finer chocolates and chocolate bars, who refined the taste by mixing it with various other ingredients. The part of the plant where cocoa beans were roasted was a high-noise area. When going through the plant I noticed a worker not wearing his ear plugs. The reason he gave was that the foreman refused to give him new plugs as replacement for the old ones that were spoiled and discarded by him. I was surprised, as the rule in the company was to have an ear-plug dispenser at the entrance to the plant, where anybody can help himself. The dispenser was empty. I went with him to the foreman and asked why the dispenser was empty, and why the worker was not allowed free access to ear-plugs.

His answer was astounding, to say the least. He said that a few of the persons in the plant used too many ear-plugs, and he had to control their use to save money. And then he went to his cupboard, opened it and took out a pair of ear-plugs (they cost only a few cents!) and gave it to the worker, warning him not to come again asking for new ones before a month was over. When I talked to the worker outside the foreman's office, he said that he and a few others always experienced this pestering at the hands of the foreman, whereas others in the same plant didn't.

Upon my mentioning this incident to the works manager later, he told me that there were two clans in the factory, each terrorising the other with minor things. He was working at bringing peace within his factory, but he was new, and the clan fights had been going on for years, and he did not know what to do about it. I suggested to him building working groups containing equal number of workers from both clans for solving safety problems, asking them to elect the chairman amongst their midst and present to you solutions of the safety problems as a team. I suggested to the works manager donating the group a free dinner for celebrating their work when the solution was delivered. Some months later he called me at the headquarters and talked with satisfaction about the improving atmosphere and less and less bickering and tension amongst the workers.

The sorry part of the story is that *he* had not noticed – before my visit – that some of the workers walked around without ear-plugs and that the dispenser was empty. Why? Because he had never carried out safety visits of the type we are talking about. As far as safety is concerned, most managers are happy to be backseat drivers, letting somebody else drive, and only criticising when they felt that the driver was not driving properly; being that, they themselves cannot control the way safety "goes" at a site.

I mentioned that at least half of the members of a working group should be workers. What constitutes the other half? It is good to have a manager take part, some supervisors and, in case of safety, the safety professionals. If the problem involves technical matters, the specialist from that branch, e.g. the electrician or the computer expert is required. Somebody from the human relations department is always helpful. And do not forget the works council or union representative.

The chairman should be elected by the group at the first meeting. It should never be the person with the highest hierarchical position in the group. The best is to let a worker or a foreman be the chairman. He can be provided with help in writing the minutes or the agenda by someone from the HR or the safety department. But the leader of the group must invariably be one of the shop-floor persons.

Later when we talk about safety committees in detail, we shall also look at working groups that are installed to solve special problems.

Where else can we involve workers? What else does management do that affects them? But of course, the operating instructions according to which they have to work. They are the ones who actually operate the machines and carry out the process steps. Their feedback for optimising operating instructions is of extreme value. See to it that your engineers involve the workers right from the beginning when they are designing the ergonomics of a new machine, or writing new operating instructions. Later on, once the machine or the process has taken up operation, they should sit regularly together to discover the kinks and unsafe conditions that have arisen in the course of the operation.

There are certain formalised methods for getting the most out of workers as a group and in the shortest possible time. The best of these is brainstorming. Let us talk about it in the next chapter.

17
Brainstorming

You must have often heard of brainstorming, and, as a matter of fact, even participated in a few sessions. Brainstorming sessions are an ideal tool for involving everybody in the solution-finding process because of its multiple advantages. Firstly, it lets everybody have a say (see the basic rules below) and thus feel they have participated in the discovery of the solution. Secondly, it brings thoughts of peers together, and this generates a chain reaction of ideas. Thirdly the *group* finds a solution to the problem without any input from you, and is hence convinced of its efficacy. It needs no further pressure from you to implement the solution reached.

Brainstorming, however, is only as good as the person who conducts it. And in our case, it is you, the manager. You need, therefore, to know and strictly follow the rules we shall be discussing below.

There was some discussion in recent years that the term "brainstorming" may be politically incorrect, as it would hurt the sensitivities of those persons who suffered from a mental disease called brainstorm that caused violent symptoms of mental disturbance. "Thought showers" was suggested as an alternative (as it was presumably not offensive to raindrops!). A British study amongst brainstorm patients showed that they did not find the term "brainstorming" offensive, and thus we can continue using it safely without fear of legal implications.

Brainstorming was invented and first introduced around 1940 by Alex F. Osborn, who was the "O" in the world's fourth largest advertising agency BBDO (Batton, Barton, Durstine & Osborn), during the time he was in charge of the company (1939–1946)[29]. He had found

29) Osborn said later (1951) that he did not actually invent it, but got the idea from the type of conference procedure used in India for more than 500 years when teachers (gurus) worked with their students. The Sanskrit name for the method is Paria-Darshan, which means looking at a thing from beyond the facts.

Managing Safety. Kishor Bhagwati

ISBN 3-527-31583-7

that the normal business meeting procedures were intimidating for junior or not very eloquent people, who mostly desisted from contributing new ideas out of fear of being considered wrong or stupid, and kept quiet. As far as creativity was concerned, he found that just those "silly" ideas often led to new ways of thinking, generating innovative solutions. The first rule Osborn coined, therefore, was the "defer judgement" rule.

The four basic rules for brainstorming originally propounded by him in his book[30) were:

1. Criticism is ruled out. Adverse judgement of ideas must be withheld till later.
2. "Free wheeling" is welcomed. The wilder the idea, the better; it is easier to tame down than think up.
3. Quantity is wanted. The greater the number of ideas, the more the likelihood of useful ideas.
4. Combination and improvement are sought. In addition to contributing ideas of their own, participants should suggest how ideas of others can be turned into better ideas; or how two or three more ideas can be joined into yet another idea.

A few words about the way it runs: a leader guides the meeting, a scribe writes down all ideas generated on a flip-chart, and the best idea(s) are selected by voting, all this within a given time limit. We shall go into details right away.

However, before we discuss the procedure, let us go a little deeper into the principles that Osborn gave us:

1. Criticism

I am sure you know what this means. All meetings have persons criticising others for their "impractical" suggestions. Mostly these are persons, who do not have anything new to contribute themselves. A famous phrase they start the critic is "You must be careful ..." Then there are people who laugh demonstratively at a new idea from a person they do not like. Even the nod of a head (with a nonspoken "Oh,

30) "Applied Imagination", Charles Scribner's Sons, New York, 1953.

God", or turning the eyes to the ceiling) or a groan can be discouraging. The senior ones tend to kill the idea by saying things like: "Oh, we already tried that before. It doesn't work."

These know-alls with their superior air are the best destroyers of new ideas, and have to be silenced in a brainstorming session immediately. Your job, as the session leader, is that of the conductor of an orchestra. You do not play an instrument yourself, but you govern the quality and the flow of music – in our case, ideas. Anything creating dissonance has to be silenced, or disciplined. But remember: You are conducting the orchestra in a public performance, and not during practice rehearsals. Hence you too have to refrain from commenting on a suggestion. Your job is to see to it that all are able to give their opinion freely. Make this rule clear to all participants before the session starts, but give them the hope that a discussion will take place near the end of the session.

2. Free-wheeling

The Merriam-Webster's dictionary defines "free-wheeling" as "not repressed or restrained" and "not bound by formal rules, procedures or guidelines". Let the people bring up the wildest ideas. Write them all down on the flip-chart, no matter what their content. Some "funny guys" will try to bring in ridiculous ideas wanting to create some laughter. Take them seriously and write down their ideas! During a certain period in my company, the people were unhappy with the personnel reforms that had to be carried out. The HR department (in those times called the Personnel Department, before it changed to Employee Relations Department and finally to Human Resources department) was responsible for communicating the ideas to the people. At these times, in every brainstorming session, the suggestion that invariably came up was: "Fire all people in the Personnel Department." And we always wrote it down. As Osborn said, it is easier to tame down a wild idea, than think one up. So promote unusual ideas, those that leave the bounds of things that seem feasible or possible.

Once in a while you have to interfere in the proceedings to prevent it leaving the course. Be sure to control a vulgar suggestion; not by rejecting it, but by asking the person to rephrase it in a civil language. Above all, see to it that the suggestions are short and succinct. Remember that we have put a time limit for the session (about half an hour). So there is no time for hearing and writing long-winded statements. If someone starts taking too much time (i.e. more than thirty seconds), interrupt him and ask him to stick to the point and try to say it in only a few words. It is very important that you make it clear at the start, and then later during the session, that this is not a debating club of a philosophical society, but a session where you expect a concrete answer by the end of the given period. Lead the people away from long explanations, and bring them back to the three or four words that describe their idea.

3. Quantity

The previous sentence is important to remember, because we want as many ideas as possible. If people take a long time to get to the point, we will have too few ideas within the time limit we have set, and the purpose of brainstorming would be lost. As a "conductor" you can control this by asking each one in quick succession, and if someone has no suggestions to make, jumping over to the next one immediately. In the next step, we will be reducing the quantity to make the voting more manageable, but for this step, just prod and ask quickly for ideas. Your job is also condensing the wordings of the ideas into short, three or four word phrases that the scribe can write quickly on the flip-chart. Before getting this done, ask the person who suggested the idea whether your condensation is acceptable to him, and whether it says what he wanted to say.

4. Refining the information

During the idea input session you will notice that a few ideas are nearly the same, but expressed differently. If you notice it at once, ask the person whether the idea already noted down reflects his idea in full, or not. If he agrees, his idea need not be written down. If he, however, thinks that his idea is slightly different from the one that is already on the chart, ask him whether he can express it differently. Do not enter into a discussion at this stage.

When the river of ideas has dried up, let all look at the points again and try to find whether some of the ideas are redundant because of duplication, or can be combined by adding a word or two to express both suggestions. For example, if one suggestion was to wear safety glasses in laboratories, and the other was to have safety glasses equipped with side shields, the suggestion could be combined by stating that safety glasses *with side shields* must be worn.

If a person suggests modifying another's suggestion to make it more effective, ask the person who suggested it in the first place whether he agrees to this modification, and if he does, you can replace the old one with the new one.

These are the principles on which brainstorming is based. Deviations from these basic rules have been suggested, but I have not found them to be much better or more effective. What you basically want is that people should think about their problems, and find the solutions to the problems themselves in a team, not awaiting the solutions to be served to them by someone higher on a silver platter.

Let us now look at the procedure of a brainstorming session.

The first and the most important thing is the statement of the problem that is to be solved. This statement has to be prepared by you. The answers you get depend on how specific you have phrased the problem. The more generalised your statement, the more vague the answers you will get, and at the end of the session, which may not be finished in the given time, you may not have anything tangible to go on. Your work, therefore, is in preparing a statement that has, and thus demands, a specific answer. For example, if your statement is: "Improving safety at the grinding machine", you will have a glut of ideas that would be difficult or impossible to condense into specific actions within the given time. However, if your statement is: "Eye-protection at the grinding machine", then you will have more specific answers.

If you want to be still more specific, put the statement as a question. "How can eyes be protected from injury during grinding at the grinding machine?" Suggested solutions to such questions could be simple, or quite complicated and expensive. For example, someone may suggest putting a complicated electronic device or fully screening off the machine. Such suggestions have two disadvantages: firstly, we are back at the engineering board, trying to find a technical solution, a route that we want to generally avoid because what we are searching is a behavioural change, and secondly, refusing an expensive solution will create the wrong impression in the minds of the people that the company does not want to spend money on safety.

So we narrow down the question further. We do not just ask for any solution, but a solution we could implement ourselves within our financial and administrative constraints. It is not shameful telling your people that you do not have an unlimited pot of gold at your disposal, but a limited budget, and that the solution you are seeking is something that is affordable by the department. The question now becomes: "How can **we** achieve (or manage) that colleagues working at the grinding machine do not hurt their eyes?" Did you notice the "we" in bold letters? It is very important to stress this again and again. The problem is a home-made one, and we have to solve it ourselves with the money we have.

Leave the question open, and do not include a solution near to your heart in it, or solutions you do not want. In the above question, if you put phrases like "by using safety glasses" or "without constructing a shield" you would destroy the purpose of brainstorming.

Before scheduling a session, the following preliminaries should be taken care of:

- Announce the date and time of the session on the notice board a few days in advance. Do not forget to mention the time at which the session will end.
- Mention the theme you want to discuss in the announcement.
- The seating arrangement should be ideally circular (or rectangular), and not in a school-bench fashion. As people do not have to write down anything, just a circular row of chairs will suffice. Due to the same reason, if you do not have a table large enough to accommodate all participants as for a normal meeting, just crowd them around the table. Try not to build two rows.
- Arrange for the equipment to record all ideas, e.g. a flip-chart, a white or black board, a computer with a large monitor or a video projector and a large screen, etc.

Now comes your part in the session:

- Begin the session by welcoming all. For the first few times you convene a session, state the ground rules at the beginning of this chapter, and the time you have allotted for the session.

- Mention that each and every one will be asked in turn to give an idea. Nobody should speak out of turn. The person whose turn it is to speak may pass at this round, but will still be asked if he has something to contribute at the next round.
- Insist that no explanations are to be given for ideas.
- State the reason why you called this meeting, i.e. why and how you became aware of the problem, e.g. "In the last six months we had three eye injuries at the grinding machine".
- Put your statement or question.

This is the end of your introduction. From now on you are just seeing to it that the rules are being kept and that discipline is maintained. You ask the first person in the round to give an idea, see to it that it is written down, then ask the person next to him, and so on. The rounds continue till the ideas are exhausted. If an idea is not expressed in a few words, help the person condense it, or condense it for him, asking him whether he agrees with your shorter statement of the idea, and whether it expresses what he wanted to say.

The scribe writes down all the ideas, numbering them sequentially. It is best to divide the page in three columns: the first narrow column for counting the votes, the second narrow column for numbering the idea and the third broad column for writing down the ideas. The scribe can also give suggestions, and has the right to vote. This, of course, only if you are not the scribe. The leader, i.e. you, does not contribute any ideas, and has no vote.

After the idea generation is complete, all look at the ideas on the board and try to see which ideas are similar and can be combined into a single statement. If you lead this exercise, you have the opportunity of combining weaker ideas into others, thereby diplomatically avoiding rejecting or dismissing the weaker ideas. This is a team exercise – so do not let a person feel bad because his idea is rejected, otherwise it will have the reverse effect on the team spirit of the rejected person.

If you combine ideas, you will have to renumber your list. Now comes the voting part.

Each person has three to five votes (depends on your choice; I always gave only three votes). Each one is asked in the round to give his votes to the ideas he thinks would help solve the problem best. Passing during voting is not allowed! The scribe notes down the votes by putting a stroke for each vote in the first column. At the end, votes are counted, and the winner determined. If there is no clear-cut winner, ideas that have only one or two votes are eliminated, and voting is done again. In the case of a tie, it may be necessary to either accept both ideas, or vote once more only amongst the two finalists.

Now that the solution is there, it has to be decided who amongst those present will take care of getting it implemented. If it has something to do with loosening the strings of the money-purse, it will be, of course, partly your job. For all other cases, ask for a volunteer to manage the implementation, and extract from him a deadline by which he feels he can get the job done.

A small announcement should be placed on the notice board describing the problem and the solution arrived at by the group, and the name of the person who has accepted the responsibility of completing the work and by which date.

The first few times you do this, it will be a slightly bumpy ride, because people are not yet fully accustomed to it. But after a few sessions, the ride is smooth and fast, and you normally need less time for it than announced. It is a wonderful tool for involving your subordinates in the decision-making process, not only limited to safety, but also for many other administrative matters. Use it extensively, and you will have a satisfied workforce producing the best results.

18
Pillar 4: Accident Investigation

Although, as a top manager, you may not be directly involved in an accident investigation, you must know its importance for accident reduction and increase in productivity. My experience has shown that this is the weakest point in nearly all the organisations I have consulted, bringing all other efforts at improving safety to a dead-end halfway to success. It has turned into such a formality and form-filling activity, that the root cause for the accident never surfaces. Additionally, all the new logical and mathematical methods together with their computer software have misled the investigators and safety specialists into believing that the real cause can be found by just following certain set methods and pressing the right keys on the computer. About 95% of accident-investigation reports I have seen have lacked the proper questions to arrive at the root cause.

Trevor Kletz, the doyen of safety communicators who worked for 38 years in ICI in the field of safety, gives a very nice example of what the finding of root cause should be like. In his book[31] he writes:

"Accident investigation is like peeling an onion or, if you prefer a more poetic metaphor, the dance of the seven veils. Beneath one layer of causes and recommendations, there are other, less superficial layers ... Very often only the outer layers are considered and thus we fail to use all the information for which we have paid the high price of an accident."

For example, there are two dentists in a town, a Dr Black and a Dr Decker. A person once had terrible toothache, so he went to Dr Black, who gave him a pain killer and told him to go home. But the pain did not subside. So he went to the dentist the next day. He gave him a stronger pain killer. That helped for a day, but the pain was getting

31) Trevor Kletz, "Learning from Accidents in Industry", Butterworths, 1988, 1990.

Managing Safety. Kishor Bhagwati

ISBN 3-527-31583-7

worse and worse. Next day he was given a still stronger pain killer, but that did not help much either. So he went to Dr Decker.

Dr Decker took an X-ray of his jaw and found an infection at the root of the painful tooth. He asked him to sit down, gave him an anaesthetic injection, and carried out root treatment to remove the infection. He was cured immediately, and was free of pain after that. Why could Dr Decker cure what Dr Black could not? Because Dr Decker looked deeper into the cause of the pain, and not just at the symptoms. If you had a choice between the doctors Black and Decker, whom would you choose? The superficial examiner Black or the thorough examiner Decker?

Unless we know the final root cause of an accident that needs to be removed, we will not be able to prevent the accident happening again. As the whole purpose of an accident investigation is to assure that the same or a similar accident will never happen again, we have to dig deeper and deeper till we find the act or condition that triggered the various steps that finally led to the accident.

> The only purpose of an accident investigation is to assure that the same or a similar accident never happens again.

The question you have to ask whenever an investigation report is presented to you is: Can the countermeasures that have been decided in this report guarantee that the same or a similar accident will never ever happen again? If you do not get a positive answer, tell the investigators to go back and search further till they can answer your question positively. It is only your insistence in this matter that will produce positive results. If you do not ask for it, as in most companies I know, the persons investigating will not put in the extra effort to find the root cause, and similar accidents will happen again and again.

The trouble with many accident investigations is that they metamorphose into an inquisition. The persons carrying out the investigation have only one purpose: To find out whom or what to blame! Although companies do not like to hear it, blame culture is very prevalent in many organisations. For every mishap somebody will be searched on whom to put the blame. This, like any other cultural direction, starts at the very top.

I once evaluated the safety-management system of a private railway company (after privatisation of British Rail) in England. One cannot imagine the extent to which the blame culture had taken hold there. Anybody you asked about something improper had always a finger pointing at someone else who was responsible for the poor condition. When I mentioned this during my debriefing to the management board, there was quite an uproar, all crying out that this was not the case. Then the CEO looked at me threateningly and said, "Who are those guys who told you that? I want their names!" Of course, he did not get their names, but his first instinct was to catch somebody and blame him.

If you start an accident investigation with the purpose of finding a culprit, you are on the wrong track for improving safety. On top of that, if you have a system of punishing the culprit, you can forget ever reaching the goal of good safety performance. The so-called culprit is mostly the person who was hurt, the victim, because he did something that was not safe. What do investigators usually do when they have decided whose fault it was, or who the culprit was? They think of what the fine should be, and close the investigation, with the result that the chances of finding the root cause are gone. This is why successful companies do not mention the name of the victim in their investigation reports. On insurance forms the name has to be mentioned, but not on company internal forms.

One of the most frequent sentences you find in the box for "Causes" in accident reports is: "Loss of attention or absent-mindedness". And the usually given countermeasure to prevent a recurrence: "The worker was told to be more careful (or attentive) in future". As mentioned in Chapter 5, this is as effective as water on a duck's back. Our own children are excellent examples of what effect all such talking has. I do not know of any worker who started becoming more careful because someone told him to. As we have seen, the reason for so-called carelessness is actually the conscious wish to take a risk. What is the use of telling him to be more careful, when the real reason behind it is not being addressed? With an approach like this accidents will continue happening to the same person or to others in the same situation.

What we have to discover, therefore, is the unsafe act that resulted in the accident, and not the unsafe person. Having discovered the unsafe act, one can start digging deeper to find out what was the trigger for that act. That trigger may again have further triggers back in chronology. We, therefore, have to dig still deeper.

There is a simple rule when investigating accidents: ask at least 5 times questions like "why?", "what?", "where?", "which?" or "how?" Here is an example I have heard in a seminar:
A man walking in the warehouse of a factory slipped on an oil patch on the floor and could not come to work the next day. This was, therefore, a recordable accident[32]. After two days the works manager received an accident-investigation report written by the safety professional as follows:

Facts: The injured person was from the internal sales office, and visited the warehouse to check on an order. While crossing the floor, he slipped on an oil patch, and was unable to come to work the next day. He returned to work the day after.

Cause of accident: Lack of attention while walking. He should have seen the oil patch and gone around it. Additionally, not being warehouse personnel, he should have stayed within the marked walkways.

Measures to prevent a recurrence: He was told to be more careful in future, always stick to the marked walkways when going through the warehouse, and not to walk in working areas.

In 95% of companies, this report would have gone unscathed and taken its final resting place in a file somewhere. The works manager here, however, was smarter. He sent the report back and asked for a deeper investigation, using the 5 W-questions. The result was as follows:

Q: *Why* was the oil patch there?
A: A fork-lift truck (FLT) was losing oil from its hydraulic system. The driver of the FLT said that he had been noticing it for three days.
Q: *What* had been done to the FLT three days ago?
A: It was repaired in the company's maintenance shop.
Q: *What* was done there?
A: The FLT driver had complained that the brakes did not work very well. Hence a valve was changed in the hydraulic line. It was a new type of valve, not the original spare part. When fitting it, it was noticed that it did not fit smoothly,

32) See Appendix 1 for recordability of accidents.

but the purchasing department had sent it with the assurance that it was equivalent to the original spare part.

Q: *Why* did the purchasing department buy a different valve?

A: The salesman who sold it had assured that it was the same as the original spare part, fulfilled the original specifications, but was 30% cheaper. And that is the job of the purchasing department, isn't it?

Q: *Why* did the purchaser not check its suitability with the maintenance shop?

A: The purchaser had enough technical knowledge to understand the specs, and besides, there was no provision in the company rules that he must do so.

Let us now start imagining the further consequences of this oil patch. One-by-one, all the FLTs would have been fitted with this new valve and oil patches from leaking FLTs would have been all over the place. There would be more slips and falls, and one day, an FLT would slip on one of them, the driver losing control of its direction, and perhaps forking a worker standing near, or crushing him to death. Never say such things never happen; they do! Many accidents have their roots in the most improbable things happening simultaneously.

Had we been satisfied with the "analysis" in the original report, we would never have been able to avoid the dire consequences described. The cause of the accident deep down was not the oil patch, but the purchasing system. Since that time, no purchaser in that company is allowed to decide whether he can replace a spare part or any other technical material, without getting approval from its user, e.g. the maintenance shop. The user is asked to check the compatibility of the part, test it, and if found OK, approve it.

One more discovery through the analysis was that the FLT driver, who had been noticing the oil patches he had been leaving behind since three days, had not reported it immediately to his superior or to the workshop. Neither had the cleaning crew working in the warehouse.

Do you think a recurrence of the accident could have been positively prevented by reprimanding the person injured to walk more carefully in future?

Accident investigation is the magic wand that prevents further accidents. In the next chapter we shall look at the methodology of proper accident investigation.

19
The Methodology of an Accident Investigation

Accident investigation is a tricky thing. If not carried out properly, it does not produce the desired results. Here are some of the basic rules for carrying out an accident investigation.

1. The investigation should start as soon as possible after the event has happened.
2. A team of three to four persons suffices.
3. The investigation is to be led by the victim's immediate superior.
4. No culprit is to be sought.
5. A one-page report is to be written.

The investigation of an accident starts with gathering information. In all the whodunits you have read, you must have recognised that 95% of the story is concerned with finding the background facts. A Sherlock Holmes or a Hercule Poirot never jumps to conclusions without first having collected all the facts. For collecting facts in case of accidents, it is necessary to question people. If the questioning technique is not proper, answers will not be forthcoming. One may even hear lies. White lies that people invent to protect themselves, like the lies policemen often hear when they stop a person driving too fast in a township where the speed limit is 50 km/h. If asked at what speed he was driving, the driver would say: "Well, 49 km/h at the most. Not a bit quicker."

Such questioning techniques have to be learned. They hardly come naturally, and do not form a part of the curriculum at schools and universities. I shall be giving some pointers for this in the following chapter. Let us first go through the points mentioned above one-by-one.

Managing Safety. Kishor Bhagwati

ISBN 3-527-31583-7

1. When to start investigating?

The answer is: as soon as possible, especially before the witnesses have had a lot of time discussing things amongst themselves. Facts become distorted with time, not necessarily willingly. The human memory is not that strong on details, and the more one thinks about it, the more the facts start accommodating themselves to a fixed perception. I always used to be astounded when Perry Mason asked the witness to describe a certain person, and the witness could shoot off: "He was about 5 feet 7 inches tall, brown hair, hazel eyes and weighed around 145 pounds." If you asked me a few hours later to describe his hair colour, I would be at a loss, to say nothing of his height or weight!

It is, therefore, absolutely necessary that the investigation is started as soon as possible. I have seen reports in companies where the investigation started a few days, or even a week later. The farther away you are in time from the event, the more the facts start becoming hazier, and the answers you get do not lead you to the root causes you are seeking. The two famous sleuths mentioned above were usually at the spot of the crime within the shortest time (Holmes having the Bradshaw in his head), or were there when the crime was committed.

2. The investigation team

In many companies, the investigation team consists of the responsible manager or (mostly) his representative, a supervisor, somebody from the HR, a representative of the works council or union, and one or two members of the safety department. Chairman of the team is the safety specialist. When some special technical problem is involved, somebody from the appropriate technical department is also requested to attend. Such teams tend to be formalistic, and to produce reports that are pages long, with diagrams, photos, etc. A lot of manpower is used up at a stage where the necessary facts are not yet at hand.

It is better to have the preliminary investigation, or in most cases also the full investigation, done by a small team consisting of the injured person's supervisor, one or two of his colleagues and, if required, a technician from the department that took care of the machinery involved in the accident. No person from the safety department is required, unless there is some problem with a safety-related issue, e.g.

exposure to hazardous substances. That's all! The accident happened in the department, and it is the department's responsibility to take care of it. A report of the investigation will be spread to all who have a need to know, and questions can then be answered, but for the present it is a departmental matter.

Most managers think it is not their, i.e. their department's responsibility, and love to pass the buck over to the specialists, the safety department and ask them to lead and carry out the investigation. After all, that's what they are there for, isn't it? They are specialists! They will analyse the problem in a professional manner, and give me the ready-made solution. I can await their response. There are so many more important things waiting for me. I can't be expected to take care of everything in my department.

Sorry, but you are wrong. Where did the accident take place? In your department. Where did the injured belong? In your department. Who is going to implement the countermeasures? Your department. Who is going to have to rearrange workplace schedules because of the absence of the injured? Your department. Who is going to have a production interruption because of the accident? Your department. And now you dare say the accident is not your department's business? Think again, sir! It is all yours, lock, stock and barrel!

3. The investigating team leader

The team is chaired by the victim's immediate superior, in most cases his supervisor. He is always involved at every step of the investigation, even when, due to the gravity of the case, the responsibility for the investigation is taken over by someone higher in the hierarchy. There are very good reasons for making him the chairman of the investigation team. Firstly, he has the responsibility to see to it that his workers do not carry out unsafe acts, a responsibility that he fulfils by frequently going around his area with an eagle eye and stopping workers from doing unsafe acts. Secondly, he knows his workers and their work better than any outside person, e.g. from the safety department. He knows their strengths and weaknesses, often has knowledge of their personal problems, and has a more intimate relationship with them than others. The workers do not feel uneasy when talking to him, because they have been doing it all the time. Thirdly, an investigation by the supervisor sends the right message to the worker regarding his commitment to safety.

That in certain major accidents, which result in maiming or even death, somebody more senior is expected to take over the rudder is quite clear. In such cases (in the case of death always), the criminal investigation department and the public prosecutor will be mostly involved. The representative of the company in these cases is clearly somebody from the top management. This will also be necessary in cases of legal disputes related to injuries (claiming negligence on the part of the company). In case you have to face such an unpleasant situation one day, it is good to know in advance how things are running in your factory and have an overview of the safety situation, something you can achieve through regularly carrying out your safety visits.

4. No culprit is to be sought

The moment a person is afraid of being made responsible for something untoward he did, he starts building up internal defences to avoid punishment. As mentioned already, these defences may consist of white lies, or fabricated scenarios that show him more as an unwilling bystander who was victimised, rather than as the instigator. The person lies to protect himself, a basic trait in the human's fight for survival. Just think of the time when you, as a tiny tot, were caught by your mother taking biscuits from the jar. If the mother asked you what you were doing, you hid the hand holding the biscuits behind your back and said with an innocent look: "Nothing!" Companies with superior safety performance, therefore, do not have any names in the investigation reports, except those of the persons involved in investigating the accident.

A culprit is usually sought for the sake of punishment. In occupational accidents we are discussing, it is almost always the victim. Do you really want to punish the poor fellow? Don't you think that through the suffering and the injury he has gone through he has been punished enough? Are you going to make a better worker out of him now by punishing him? Do you think that because he has been punished he will work more safely in future? Do you want your workers to work safely out of fear of punishment? What type of work climate do you want to create?

When giving my presentation for a consulting job to the board of a company, one of the board members repeatedly grunted during my speech till he could not hold it out any longer and burst out harrumphing: "I don't think we need any consultant to improve our safety. Thousands of people, highly knowledgeable in their field, perhaps more knowledgeable than you, have put up rules and regulations on working safely. Our problem is we do not make our workers follow those rules strictly. Some iron-handed discipline with a few exemplary cases of painful punishments would do the trick. All this convincing business is just nonsense!"

I was taken aback at this outburst, but seeing the smile on the faces of other members, gained enough confidence to ask him whether he had any practical instances he could quote, where such an approach had led to lower accident figures. Of course, he did not, but that was the trouble with the world anyway. I further wanted to ask him if he had children, and whether he had educated his children in this manner, never sparing the rod. Fearing that his answer would be positive, and not wanting to antagonise a board member at my acquisition presentation, I did not say anything. Fortunately, the CEO interrupted the exchange and asked me to continue with my slides.

We already discussed the disadvantages of a blame culture. Finding a culprit is nothing else but finding someone to blame. And, as the example with the oil patch showed, having found the person to blame, the investigation stops, and valuable information that could have avoided a repetition of the same or a similar accident is lost. Therefore, right from the beginning, we have to make it clear that the purpose of the investigation is to discover the unsafe act or condition that led to the accident, and not to find the person to blame, and that nobody will be punished, whatever faulty action he may have done. As Trevor Kletz says in his book (loc. cit.): "An accident may show that someone does not have the ability to carry out a particular job and he may have to be moved, but this is not punishment and should not be made to look like a punishment."

Just *saying* that the investigation is not being carried out to find someone to blame does not mean that people will accept it as the truth. It will take some time and quite a few investigations proving this before people are really convinced that the company means what it says. As a manager, you can always check whether this is being practiced in your department or factory, by scrutinising the accident-investigation reports you receive. If you ever see a tinge of this type of approach in a report, immediately phone the supervisor leading the investigation and ask him to change his method.

5. The investigation report

This is where specialists have a field day. I have seen so many unreadable (and unread) ten- to fifteen-page reports with diagrams and pictures, nowadays even in colour, which never reached the first base in finding the root cause. All one needs to express the thoughts we have discussed until now can be put on one, or maximum two page(s). An accident-investigation report is not a newspaper report or a running commentary of the accident, but the place to put the root causes and the countermeasures the team has arrived at to prevent a recurrence. When (the still immature) Boris Becker was interviewed when he first won Wimbledon singles in 1985 at the age of 17, it sounded like this: "Well, I first took the forehand to the left, then the backhand when he shot back, then again the forehand nearer to the net, and then he gave me a volley which I then smashed onto his side of the court." Nice to hear all that, but in our accident reports we just want to know how the match went (for your information, 3:1 for Becker against Kevin Curren).

Remember, we are talking of personal injuries, and not of mechanical accidents such as equipment breakdowns or fires and explosions. In such cases extensive documentation is necessary, not only for investigative purposes but also for legal and insurance purposes. The same is valid for serious injuries or deaths. However, for the managers to grasp what has happened in such cases, it would be good if the report writer made a one-page summary of salient points and conclusions. *That* report at least will be read, whereas the extensive reports will have to wait their turn "when there is some time." The great majority of accidents in the industry, fortunately, do not fall in the serious categories, and there the one-page rule applies.

On page 148 you can see the type of one-page form that would suffice for most injuries. The form has two purposes: one is to force the people to stick to the basics, and the second is to inform others within and outside the department of the results of the investigation. The purpose of putting the report on the notice board is to activate other workers to think of the situation in which they themselves could come, and what to avoid in such cases. Also, because the report will contain names[33] of those responsible for implementing the countermeasures and the (self-imposed) deadline by which they are expected to complete it, that person will be reminded by his colleagues to keep the deadline, and the manager need not constantly follow him up to check the completion of the measure.

The distribution is to be done prudently, sending only to those who may have a learning effect from the report and may profit from it. For example, if you had an accident in your department with a certain machine that you knew also existed elsewhere in the organisation, you should send a copy to those departments or sites to make the people there aware of the hazard associated with the machine. Even the mechanical improvements you may have done to the machine to make it safer could be profitably used by other sites. There is no secrecy in safety. The knowledge gained is for all to use.

What one should not do is to distribute it all over the organisation to show what you have done and how good you are. Once a young engineer I had sent to a course in dynamic something or other came back from the course, and distributed the handout material of 67 pages to about 50 persons in the factory. Were those pages impressive! Triple integral signs decorated the formulae that were like three-storied buildings, exponential signs were all over the place, very high mathematics, I am sure. Anybody who could understand it must be a very intelligent person! What luck we had him in our midst! Unfortunately, he never got any response to his distribution, and nobody told him how smart he was, because nobody had or could read it. Quite disappointing for him.

Once I had been in the office of the president of my company, when his secretary entered with a big pile of paper about 10 cm high, wordlessly replaced a similar pile on the left upper corner of his table, and went out. I asked him what it was. He sighed and said these were the monthly reports from all the 27 sites we had in our geographical division (Europe, Middle East, and Africa). So I asked him what he did with them. He said they rested there unseen till the beginning of next

33) Whenever listing names of people of mixed positions, never go according to the hierarchy, but always strictly alphabetically. This is also a part of respecting the worker's opinion and input by not putting him always at the bottom of the list.

<table>
<tr><td colspan="2">Accident Investigation Report</td><td colspan="2">Plant:</td><td colspan="2">Section:</td></tr>
<tr><td colspan="6">Accident No.</td></tr>
<tr><td>Date:</td><td colspan="2">Time:</td><td colspan="3">Weekday:</td></tr>
<tr><td>Job of injured person</td><td colspan="5"></td></tr>
<tr><td>Type of injury</td><td colspan="5"></td></tr>
<tr><td>Treated where</td><td colspan="5"></td></tr>
<tr><td>Expected days of absence</td><td colspan="2"></td><td colspan="3">Recordable?</td></tr>
<tr><td colspan="3">Preliminary information distributed on:</td><td colspan="3">at:</td></tr>
<tr><td colspan="6">What happened</td></tr>
<tr><td colspan="6">Cause of accident</td></tr>
<tr><td colspan="4">Countermeasures</td><td>Person</td><td>Deadline</td></tr>
<tr><td colspan="4"></td><td></td><td></td></tr>
<tr><td colspan="6">Investigation Team (person, position)</td></tr>
<tr><td colspan="2">Start of investigation</td><td colspan="2">End of investigation</td><td colspan="2">This report date</td></tr>
<tr><td colspan="2"></td><td colspan="2"></td><td colspan="2"></td></tr>
<tr><td colspan="6">Distribution: Works Manager, Plant Manager, Plant Safety Person, Central Safety Department, Notice Board</td></tr>
</table>

month, when the secretary replaced them with new reports. He did not have time to read, or even glance through all of them. And if there were anything important or urgent requiring his attention or decision, he would get a call or an e-mail about it. At that time he can ask the caller or sender for details. So why waste time now?

Believe me; nobody likes to read long detailed reports – not even you. See to it that the reports of accident investigations written in your organisation are limited to one or two pages. If you get a 10-page report, send it back to the sender asking for a one-page summary. Introduce a form similar to the one given here to educate people in being brief and to the point. We shall be talking about communication within the company later, but let it be said here that anyone needing ten pages to write an accident-investigation report has loaded himself with too much ballast of facts, but has not crystallised the essence out of them. A German proverb says: "*In der Kürze liegt die Würze*", which means the spice (the interesting part) is in the (communication's) brevity. Tell your people to take this to heart because that is the way you want it.

So what is your role as a top manager in the accident-investigation process? You have the power to steer it in the right direction. Your tools are listed below:

- Let all accidents be reported to you within 24 hours of their happening. "Reported" should mean that you have to have it in your hands, or have knowledge of it. In cases of serious injuries, you have to be reached wherever you are and informed immediately.
- React to the information soonest, best by calling the senior most person under you responsible for the department in which the accident happened. He had better be in a position to give you the details you ask for.
- Before asking any other question, ask about the health of the injured. If he has been hospitalized, get flowers sent to him with your note, and if it is very serious, take time to visit him in the hospital within a day or two. You cannot imagine how happy you would make him, and what wonders it works.
- If it is a maiming accident or a death, it is your duty to go to the family and comfort them and promise them any assistance the company can give. Do not delegate this task to anybody else. It goes without saying that you will attend the funeral.
- If it is a case where the authorities may come in, involve yourself in the proceedings and take over the responsibility for the factory.
- In no case should you lose interest in further developments. Keep on asking frequently about the health of the person and the progress of the investigation.

- Read the accident report under the criteria we have discussed in this chapter, and call in the person who wrote it. This is the opportunity to ask the critical question: "Would the countermeasures proposed in this report guarantee that the same or a similar accident would not happen again?" If the answer is not a definite yes, "back to the bowler".
- During your next safety round, visit the injured person if he has started work again, and spend some time with him asking him how he is, and what he suggests to prevent a similar injury to someone else. Make a point of visiting him during the next two rounds also.

Just as with the safety visit reports, Appendix 4 shows you two completed accident investigation report forms, this time of the same accident. Again you are asked to judge for yourself which team did the investigation better and in line with what we have discussed. Is it the team with Mr P Smith in it, or the team with Mr B Wooster on it? Whose countermeasures would better fulfil the demand that the same or a similar accident shall not happen again?

The next chapter describes the art of questioning when carrying out accident investigations. It is meant for those who do the questioning, i.e. your supervisors and plant managers. But you too will benefit by reading it, because the technique described can be used in many other situations.

Armed with all these points, you are sure to bring your accident investigation procedure to a level where it works in an optimum manner at preventing accidents, and raise the safety standard in your organisation. Do not let your interest flag or wane. The way to demonstrate it is by the power of question. Always ask, ask, and ask about safety matters. Never let safety matters drift into oblivion. It does not cost you anything, but keeps the flame alive and leads others to become more safety conscious. This is what a true leader does!

20
The Art of Questioning

To gain background information on accidents, one has to ask people who might know something about the circumstances, the first of them being the person injured (if this is possible.) As already indicated, the purpose of questioning is not to find out whose fault it was, so that the blame could be put on that person. This must be made very clear right from the beginning. The purpose of our questioning is to know what went wrong, and why? We are asking questions to prevent the same or a similar accident happening again. Remember our goal: we want to be able to make recommendations for countermeasures that would prevent a recurrence.

A police or official inquiry does not follow this purpose. There the aim is to find whom to blame, so that the courts can take action against the person. If at any time during the interview you feel you are slipping into this mode of a police inquiry, stop and reformulate your question. After an accident, which has disrupted the work or spoiled the safety statistics, the wish to accuse somebody and offload one's frustrations on him is quite strong. I have often heard sentences like: "That idiot! Why did he have to handle the thing without gloves?" or "Where was the mind of that guy when he switched on the machine?"

Without practical teaching, it is difficult for someone to learn how to interview properly. Each situation is different and needs a different approach. I shall try to put down a few basic rules that must be followed if the interview is to produce maximum information. These basic rules are:

- Have and show sympathy for his suffering. Do not just disregard it. It has been a shock for him, and has upset him and his colleagues. Show him that you also share his sorrow. Therefore, do not jump right into the questioning mode, but talk to him a little about how

Managing Safety. Kishor Bhagwati

ISBN 3-527-31583-7

he feels now, how was the pain, was he happy with the treatment, and what his family felt about it.

- Do not intimidate the person you are interviewing, but put him at ease. He is very probably nervous or upset. Be kind to him. If you had gone through a similar experience, you too would be similarly upset. Regard him as the source of information that is going to improve your plant's safety, and seek his help. A person intimidated is a person who will be afraid of punishment, and you lose his cooperation from that moment on. This is also one of the reasons we should keep the investigating team small, consisting of persons well known to him. A worker is more likely to be honest and talkative when he is amongst his friends and acquaintances, rather than when there are strangers present.
- Explain to him the real reason for the interview. You cannot repeat often enough that you are not seeking someone to blame, neither him nor his colleague, and that the purpose of your conversation is to find out what the trigger for the accident was. Right words could be: "I am sure you will help us in seeing to it that such an accident does not happen again to a colleague of yours"
- Do not push him to obtain the information you need. In other words, be patient. Let him unroll the story in his own way. It may take longer, but you will get so much secondary information that could be useful. Do not show your impatience by drumming your fingers or moving around uneasily in your chair. I know this is quite difficult, especially when some people talk of things you find totally unrelated to the matter at hand. But whatever he says may contain a key to what you are searching.
- Do not interrupt frequently when he is speaking. Many workers have difficulties expressing themselves, and interruptions will break their train of thought and require them to marshal the thoughts anew. Constant interruption is also a sign of the interviewer's impatience.
- Do not ask questions that contain the answers you want to hear (leading questions). With such questions you lead the person you are interviewing into the channel you have preselected. It is readily accepted by the worker, because it means less work for him. You will thus be approaching a preconceived solution, and not the true solution.

- Refrain from prompting the person you are interviewing. Do not put words in his mouth, even when he is groping for the right word. Your prompting will make him dependant on you for hints, and he will try to adjust his information to what you may want to hear.
- Encourage him to talk. Nod confirmatively when he is talking. Jog him along with short words like "Yes", "Hmmm", etc., and positive facial expressions.
- Confirm by paraphrasing his statement that you have really listened to what he said. Do not parrot his words, but use your own, different words. For example, if he said: "I isolated the machine before opening it", you say "You mean you switched off the current first?", and not "Aha, you isolated the machine before opening it?" This also serves to show him that you have understood what he said. If not, he can always say "No, I mean I removed the fuse in the panel first." In this way there is no ambiguity in the facts.
- If you have the feeling that the worker is trying to hide something, ask him to make a sketch of the situation. As compared to talking, it is very difficult to "lie" in a sketch.
- If you would like to make some notes, keep them short. You are not writing down what he says word-by-word to be able to use it against him later. So do not give that impression to him. Make only notes on some facts. In no case should you use a recording machine! These are police methods, and we do not want to use them.
- Your questions should be open-ended, and not answerable by a simple "Yes" or "No". For example, do not ask questions like: "Did you go to the doctor?", but rather "What did you do after you were hurt?" Leading questions are also not open-ended, as they too can be answered mostly by a "Yes" or a "No".
- And above all, be the great listener. Let him do as much of the talking as possible, with you jogging him along patiently till he has had his say.

Witnesses to an accident also need to be interviewed during the fact-finding exercise. The basic rules given above are valid there too. However, for witnesses we may have a list of questions that would help us further. They are less prone to feel that they will be blamed for the accident, as they had only witnessed the accident and were not directly involved in it. Still you have to be careful, because some witnesses may try to protect their colleague or to avoid embarrassment, by lying a little. What is more important, however, is the natural tendency of humans to adjust facts to a logical sequence, and eliminate all *nonsequiturs* to make the event as clear and logical as possible. This leads to unintentional twisting of facts till they fit. The more time you allow to let this happen, the less truthful the account you will get. The only way to avoid this is by taking the interviews as soon as possible after the accident has happened.

Below is a list of questions you can put to witnesses:

- Where were you at the time of the accident?
- What were you doing at the time?
- What did you see, hear?
- What was (were) the injured worker(s) doing at the time?
- In your opinion, what caused the accident?
- How can similar accidents be prevented in the future?

The figure below gives in a tabular form the Dos and Don'ts of interviewing we discussed above.

When interviewing

DO...	DO NOT...
• Put the witness at ease • Clarify the real reason for the interview, e.g. to find facts, and not culprits • Let the witness talk as long as he wants • Be the great listener • Ask open-ended questions • Make very short notes • Rephrase his answer to show him that you have listened and understood	• Intimidate him • Give him the feeling that you are searching someone to blame • Interrupt him often • Show your impatience • Ask leading questions • Make detailed notes or tape-record • Repeat just word-by-word what he has said

This chapter was meant to develop your skill and let you know how pitfalls in interviewing techniques can be avoided. As I mentioned at the beginning of the last chapter, you may not be the one who is investigating the accident, but you have to know when it is done right. In major cases like serious injuries perhaps you should involve yourself in the investigation, albeit as a participant, and not as a chairman. Even there you would need to know whether the investigation is running along proper tracks, or not.

Your job mainly is asking for all reports to be sent to you, reading them and questioning the countermeasures suggested in their ability to permanently remove the same or a similar accident in your area of management.

21
Accident or Incident?

If you have missed the word "incident" until now in this book, there is a reason for it. Although, etymologically, there is no difference in the meaning of the two words[34)], customary usage in the industry has made incident the overall term, and accident a subordinate term to it. Thus an incident is understood as:

- An accident with human injury
- An accident without human injury, so-called near-miss
- A potential accident

The main difference between these is:

- An accident with human injury will cost money,
- An accident without human injury may cost money and
- A potential accident may cost money in future.

A near-miss is a question of luck. Under unlucky circumstances, a near-miss (in some companies it is called "a dangerous occurrence" or "a dangerous incident") could have been an accident, even a serious one. For example, the huge chandelier in the foyer in an opera house falls down. If it happens during the time the opera is in full swing, then nobody would be in the foyer, and nobody would have been hurt. But if it had happened during the interval when the foyer is full of people with their champagne glasses in their hands, the result would have been catastrophic.

34) Both are derived from the Latin verb cadere = to fall. The prefix "ad-", which is written as "ac-" before words starting with c, k and q, indicates motion, e.g. "towards". The prefix "in-" also means "into" or "towards".

Managing Safety. Kishor Bhagwati

ISBN 3-527-31583-7

Because of this, a near-miss is to be investigated with the same effort and meticulousness as an accident with human injury. No one can guarantee that a mishap will not happen at the "right" time.

Well said! But the problem is, near-misses do not get the type of notice they should, although they point very clearly towards situations and acts that would, at some other time, result in an accident or even a death! Because they are not as visible as an accident with injury and an accident claim is not involved, they are hardly ever given sufficient attention.

Another problem with near-misses is that they are normally not recorded or reported, unless the damage is of a magnitude that cannot be hidden. Most of the time they are "swept under the carpet". This being the case, how can we gain the knowledge that would have helped us avoid an accident under similar circumstances in future? As a matter of fact, we should be happy that, in spite of conditions or acts that were pregnant with an accident, nobody was hurt. What better opportunity to search for the root of the matter?

There is only one way out of this, and the key to it rests with the managers. You must make it very clear that you want all near-misses to be recorded and analysed. If you hear of a near- miss that was not reported, make a fuss about it and make it difficult for the person who should have been responsible for reporting it. You can involve your safety department to play the detective in such cases. As Trevor Kletz explains, "Stories picked up from the grapevine can be followed up; fires can be detected from the fire service reports, spillages from the cleaning gang's worksheets; claims for damaged clothing can be probed to see if the incident concerned has been reported. In these and other ways managers can bring dangerous occurrences out into the light of day."[35)]

When carrying out an in-depth analysis of accidents, you will see that many of them had precedents that did not lead to an injury. In our drive to catch the accidents at their source and nip them in the bud, we need to know about these near-misses, i.e. times when the worker just had a narrow escape. One more reason for insisting that these must be recorded and reported to the safety department.

The third type of incidents concern the future. People get used to unsafe acts and conditions and do not notice them. For example, an

35) Trevor Kletz, loc. cit.

uneven tile in the flooring, which could trip a person, goes unnoticed for months and years, because people working in the plant know that it is there, and always avoid it. That is, till someone with a load in his hands stumbles over it and hurts himself. It is not that the people in the plant have not noticed it, it is just that they have got used to it being there. They always expect someone else to do something about it, and as long as that does not happen, they find a way of going around it.

I once visited the control room of one of our plants. To enter it, one had to push a swing door. I pushed against it with normal pressure and walked in. However, the door jammed on the ground and stopped. I banged with my face against it and hurt myself. Fortunately, not very seriously. It could have been worse, if my glasses had broken, and a splitter would have gone into my eye. Upon asking I found that everybody in the plant knew that that door jammed and that it required extra pressure to open. Only outsiders like me or the contractors did not know it, and I was not the first one who had banged his head against the door. These are situations that need to be remedied as soon as one becomes aware of them. The problem is: becoming aware of them.

In the company I worked for, every plant was expected to produce 10 such potential safety accidents per month, and remove the potential. Although the company was in the top most rank of safety in the world, every plant was able to fulfil this requirement. For example they would find a cold water pipe whose insulation had been removed for repairs dripping its condensate water on an electrical socket below it, or an apprentice using a screw driver as a chisel to remove a bearing from a shaft. The majority of discoveries were of the mechanical type, i.e. unsafe conditions that could have led to accidents. I was glad to see that the door to the control room had at last been taken up on the list and was remedied within one day. Investigations for such potential accidents were carried out twice a month in a small team with varying members, and solutions decided were delegated to the team members for completion by their self-committed deadlines. A follow-up was done during the next round.

I recommend introducing this system at your site. You need not look into the minutes of these meetings, but a copy must be sent to the safety department. They can report to you whether the plants are submitting their reports regularly, or not. If a plant does not come up with the required number of reports for some time, they are not necessarily neglecting their duty, and need not be taken to task. It could also be a good sign, because the plant has no potentials for accidents! However, if this goes on for some time, the safety professional should look into it and offer his services during one of the rounds.

Asking people to report a certain fixed number of potential safety accidents can also backfire, as the following example shows. An oil boring company I consulted said that their customer, an oil giant, demanded from them at least 100 so-called safety observation cards a month. They had no choice but to submit them. I asked where they found all those safety-related problems. He very coolly said: "At my desk." At least 70% of the cards were fabricated in the office and submitted, knowing well that the agent of the oil company just counted the number, but never looked at them. Bureaucracy knows its own solutions that have kept it alive aeons after aeons!

22
Responsibility & Authority

Companies like to say: "Everybody is responsible for safety." You see this slogan in policies and in posters pasted all over the factory. However, if you ask a manager who this "everybody" is, he would explain that this "everybody" is the worker and the safety department. In Chapter 9 we read that management is responsible for safety. More specifically, management is responsible for *managing* safety. What is meant by that is that managers are responsible for creating a climate, where everyone is aware of the importance of safety, and knows that he is responsible for fulfilling its objectives.

Just as one does not want managers to think that the safety department is responsible for safety, one should not let the worker, who reads the sentence "Managers are responsible for safety" believe that not he, but only the management is now responsible for safety, and that he does not have to do anything, but let the managers take care of it. This sentence, therefore, has to be explained very clearly to everybody. All are in the same boat as far as overall safety is concerned, and this washing of hands in the style of Pontius Pilate is not to be permitted; neither by the managers nor by the workers.

General statements have this weakness of being misinterpreted, because everyone can interpret them as suits him best. Let us make the general sentence "Everybody is responsible for safety" more concrete. We can say:

"Every employee in this company is responsible for his own safety and the safety of any other person present in his area of work. He is responsible for seeing to it that neither he nor the other person gets hurt. This also means that he is responsible for seeing to it that he and the other person are adequately protected with protective equipment against the hazards of injury present in his area."

Managing Safety. Kishor Bhagwati

ISBN 3-527-31583-7

That is quite a lot of responsibility. To fulfil it, however, he would need authority. Authority to stop an inadequately protected person from entering the hazardous area, or asking that person to leave the area, and stopping work by others he thinks could hurt him or others present. We, therefore, continue:

"Every employee is hereby authorised to refuse work which could, with great probability, injure him or others, to stop a person not wearing the required safety equipment from working or entering his area of work, irrespective of the rank of the person, and to politely ask those that refuse to work safely to leave his area. He has the full backing of his superiors and the company for all such moves."

If, for example, the worker is a turner, the area of about one metre around his lathe is an area where cuttings could fly and hurt persons standing there. He, therefore, is responsible for seeing to it that any person entering this area, i.e. "his area", has the proper protective equipment, e.g. safety glasses. If a person, say his superior, wants to talk to him and comes nearer to his lathe without glasses, he either requests him to put on the glasses, or turns off the lathe.

Once in my company, we were expecting the visit of the Minister of Environment to see our new plant that had been built to the highest standards of environmental protection. The minister was accompanied by his permanent secretary. Our side was represented by the works manager, the country manager, our public relations person and me. We were just nearing the gate of the plant, with the minister rushing ahead and us trailing behind him, when an operator of the plant, who was standing by the gate, approached the minister with his palms open towards him, asked him to stop and said: "I am sorry, sir, but I cannot allow you to enter the plant."

The minister was surprised and asked him why. He said: "You are not wearing safety glasses, and you have not tucked your tie into the shirt, sir. We have hazardous chemicals here and rotating machines and you could get badly hurt if you went in like this." The minister, who was issued the glasses but had not yet put them on, immediately wore them and tucked his tie into his shirt. The operator thanked him and stepped aside.

Later, during lunch, the permanent secretary asked our country manager quite angrily, what type of disrespectful behaviour we had in our factory. “A high-ranking minister comes to visit your plant, is stopped from entering by a blue collar worker, and none of you managers opened your mouth and told him to behave himself! You should have shut him off immediately before he could insult the minister.” Our country manager smiled and said: “But we have given him the responsibility to see to it that nobody in “his” area gets hurt. We have fully authorised him not to let anybody enter “his” area without wearing appropriate safety equipment. He did just the right thing. He would do the same with our company President, or the Emperor of China!”

We cannot expect our workers to behave safely if we let others walk around or work in the same area in an unsafe manner. Safety rules are valid for everybody. There are no exceptions. Sometimes laboratory people, who come into the plant to take a sample, believe they do not have to wear the safety equipment prescribed. After all, they are only there for a short time. They do not want to wear safety shoes or glasses or appropriate clothing for this short period. Such arguments are also heard from other contractors, e.g. the ones who replenish the drinks vending machines. Our answer should be quite clear. “Sorry, no exceptions. If you want to enter the area, you have to stick to the safety rules.” If one allows people to obey safety rules only when they think necessary, then one better not write them at all. Would you look away and allow a VIP to smoke in a petroleum refinery?

I was once walking with a plant manager in a steelworks. It was in one of those huge plants where cars could drive in. A Volkswagen Combi of a courier agency stopped a little distance ahead of us and a young man jumped out of it with an envelope in his hand. He asked a worker standing there if he knew who the plant manager was. The safety rules for this plant required wearing a helmet and safety shoes. He had neither of them on. His feet were “protected” by open sandals. We soon came abreast of them and saw these two arguing. Upon asking, the worker explained that he had asked the young man to step back into his car and drive away, and come again only after he had the proper equipment on.

The courier complained that it was an urgent delivery for the plant manager, and he had stepped out of the car only to deliver the letter and get the plant manager’s signature. He had no equipment in the car, because he had forgotten them at the town office 6 km away, where he would now have to drive back to get it. Can’t he just give the letter to the plant manager now? His boss would be quite angry at him if he wasted time and petrol for the return trip.

The plant manager fully supported his worker's decision, refused to take the letter and told him to go back to town and wear the proper equipment as was agreed in the contract between the steelworks and the courier company. He would prefer to wait for the urgent mail, rather than allow the breach of a safety rule. Later he wrote an e-mail to the purchasing department asking them to stress this point once more to the suppliers. After this incidence, the manager of the courier company took care to see that none of his employees went to the steelworks without carrying the protective equipment with them.

Giving workers this authority makes them more safety conscious. If they feel themselves as watchmen of safety, they cannot very well afford to be sloppy themselves. It is like the police car breaking the speed limit (without its siren blaring). See how they stick exactly to the speed limit when they drive in front of you, frustrating your itching foot that wants to press the accelerator a little further. Authority also generates pride. Give a person a responsibility and the authority that goes with it, and he will carry his head higher and do things you would not expect of him in the normal case.

A teacher we know was studying for her B Ed degree, and she was sent to a school in a rough neighbourhood to give a class. Her classmates pitied her, describing how another student had left the class half-way through with tears in his eyes. She entered the class of about 40 children around 9 to 10 years of age, fighting among themselves, running around the class, jumping up and down the benches, in short a completely chaotic classroom. Nobody took any notice of her. She stood in the door, observed the children and found a leader among them, a larger boy who seemed to be the boss in the class. She called him to her and told him: "Look, I want to hold a class here, but I want your help for it. I want you to be my sheriff and bring some order in the classroom. Could you help me?" The boy's eyes brightened, his chest bulged, and he went around the class hitting others left and right and forcing them all to sit down at their places. He sat down on the first bench, beamed at her and said: "You can now start, Miss." During the lesson he always looked back and surveyed the class to see if anyone was being noisy. He got his position of importance in the class and our friend could finish her lesson in peace.

As a lecturer on industrial acoustics at the University of Bremen, I had a class of about 40 final-semester engineering students. The students were normally quite docile, but there was one student who used to disrupt the lecture often. He would try to joke and produce laughter in the class, try to show how little importance he found in what I was teaching by unpacking his sandwich and munching on it and drinking from a bottle when I spoke. I could not tell him to leave the class like in a school, or put a dunce cap on him. He really used to disturb the class and enjoyed doing it. Let us call him Peter.

I once wanted to show the students a film on how ears are affected by noise. This was before one had videos everywhere, and I only had a 16-mm film reel. Not being on the regular staff of the college but an outside lecturer, I did not know how to organise the showing. So I called Peter after the lecture and told him, "Look Peter, I have a problem and need your help. You look like someone who can help me out. I want to show a film next week, but do not know how and where to organise the projector. Can you do this for me?" He was more than willing to do it. At my next lecture, he had already set up the projector, and showed it to me proudly. I thanked him for it, but confessed I did not know how to spool the film through the machine. He beamed and told me I need not worry about it, he would stand at the projector and do all things necessary. I started my lecture thanking him in front of all, and said that without his help, I would not have been able to show the very informative film.

He did everything professionally, and at the end of the lecture gave the reel back to me, and said if I wanted anything else, I just need ask him. After that day, he always greeted me when I came to the class, never caused any disturbance and was a model student. I gave him the importance he wanted by trusting him and making him "my assistant".

Have confidence in your people and give them the importance they deserve, and you will see how well things run. At one of the plants in my company manufacturing an organic chemical, the process was run by blue-collar operators, but the monitoring of components of each intermediate reaction step was done by people from the central laboratory. Someone in a white coat would come from the lab to the plant, go to the sample point, withdraw a sample, go back to the lab, analyse it and advise the plant whether the process was running properly and within specifications. We investigated what types of analyses were done and found that they were done with automatic analysing machines (gas chromatographs), which did not need much knowledge or skill to run, not more than injecting a sample into the machine and reading the printed results that came out.

We thought that type of work could also be done by the operators themselves. So we installed two gas chromatographs in the plant, taught the operators to how to operate them and evaluate the results. This did not take long, and soon the workers were controlling their process completely by themselves. And they were very proud of it! They felt recognised not only for their labour, but also for their intelligence. Give a person responsibility, and his stature rises in his eyes. Our neighbour's daughter, who was studying medicine at Vienna in the second year, called him once and said excitedly, "Dad, today I was allowed by the doctor to draw blood from the vein of a patient *myself*. He had told me yesterday that I may do it today, and today I really did do it. Just like a *real* doctor! I could not sleep the whole night because I was so excited! He really trusted me. I am so proud now."

We already noted that the workers working for us are a chosen group, selected on the basis of their abilities, the impression they created on us during the interview, and their intelligence. If we try to control everything, we are missing the chance of utilising this huge resource we have at our disposal. Give them more authority and watch the wonder develop. Some managers keep on forgetting that even workers are human and have feelings. If you hurt their feelings, they stop becoming your helping hand.

A new buzzword is making rounds. It is "empowerment". It is something more than what we are talking about. In industry, it means handing over all the power to the workers. A factory that has introduced worker empowerment has only two hierarchical levels: the works manager and the workers. Teams of workers take care of administrative tasks, order the raw materials, check the quality, do the product handling, etc.

The word is new, but not the concept. It is more than 150 years old, and was implemented in the western world during the 1850s gold rush in California. It is based on the Chinese system of worker responsibility, which was found to be very efficient by the railroad companies who were building the railroad during this period from Sacramento in California to the Sierra Mountains. They engaged Chinese immigrants to work according to their own way, giving full control of the project to front-line work teams. This was in total contrast to the way work was being done at that time in Eastern United States, where command and control was the rule of the day.

The Chinese work teams were extremely efficient and achieved a record of laying 10 miles of track in one day on 28 April 1869, something unheard of until then. Leaders from the East Coast had to swallow the pride built on social prejudice, and ask workers for information and advice. This heightened the intensity of motivation.[36)]

Although one may not want to hand over everything to workers, the principle of asking them to supervise safety on our behalf and to provide us with advice in this field still is one of the best things you can do to improve safety.

36) The Chinese immigrants on the West Coast were so efficient that laws were passed that made it illegal for them to seek jobs or enter into businesses that Americans wanted. They could only do businesses like laundering, which were neglected by Americans. This is how the American term "Chinese Laundry" came about. Source for all the information above: Robert L Webb, "Empowerment History".

23 The Safety Committees

I do not know of any company that does not have site safety committees. Some of them have even more than one subcommittee. In many companies the safety function is combined with the environmental and the industrial hygiene function (the latter being also called "occupational health" or just "health".) Such companies have, therefore, Safety, Health & Environmental Committees (SHE or EHS or HSE). The function of the committee is hardly ever defined properly. It exists because it must, either because legislation demands it, or because other companies have one too.

Even companies with the worst safety statistics have such safety committees. They also have policies, posters, slogans, safety literature, and various other things. Why then is their accident rate high? A safety committee is supposed to be the body that steers the activities in a company in such a way that the company has hardly any accidents. Why does this not work in such companies? Why are safety committees so ineffective in many companies? What is it that they lack? I have suffered through some safety committee meetings that were like a debating club, with participants trying to torpedo each other's ideas without contributing anything positive themselves.

The (US) National Safety Council in its book "Accident Prevention Manual for Industrial Operations"[37] states: "Some organizations prefer other types of employee participation to formal safety committees because they feel safety committees require a disproportionate amount of administrative time, that they generally tend to pass the buck, that they sometimes stir up more trouble than they are worth, and that some supervisors try to unload their responsibilities onto the safety committee. The answer to these objections is not to abolish the

37) Accident Prevention Manual for Industrial Operations, Volume 1: Administration and Programs, published by the National Safety Council, Chicago.

Managing Safety. Kishor Bhagwati

ISBN 3-527-31583-7

committees but rather to re-examine their duties, responsibilities, and methods of operations."

Following are the salient points of an *ineffective* safety committee:

Such a committee
- has as members only supervisors and some junior managers
- has people only from the production departments
- is chaired by the safety professional
- does not rotate membership every year
- has no representative of the works council or union
- meets irregularly and postpones meetings often
- has sporadic attendance by members
- managers are often represented by someone junior in their department

A safety committee like this will be nothing else but the fulfilment of a formality. Not that they will not produce paper! You will get extensive minutes of the meeting with lots of points that sound as if they were solutions to matters of life and death. But at the end of the day, they will not have done anything to stimulate and maintain interest in safety in the workforce, which should be their main task.

Safety Committees are to be installed at all hierarchical levels. At the top most level is the Central Safety Committee, or the Safety Steering Committee. Further committees are at departmental levels, plant levels and section levels. The Central committee deals with factory-wide problems, whereas the lower committees deal with issues in their own area of management. Committee meetings are also very useful for up-down communications. The messages of top management, as interpreted and explained further by managers in their own area, are much more effective and acceptable than notices and flyers.

The principle of second- and third-level safety meetings is the one we already discussed, namely the involvement of workers in the safety effort. We want workers to feel that they are a part of the total effort of the company, and that they can contribute to the company's success by being actively interested in company matters, rather than being just passive recipients of second- or third-hand information. If we want to achieve this, we have to provide for it. For example if there is a mishap in the company, we have to inform our own workers immediately, giving them as much information as we have at hand. Outside the factory, the worker will be asked by others what happened, and he would look quite silly if he just had to shrug and say he did not know, because nobody told him in spite of his being a company employee. On the other hand, if you gave him the information before he left the plant, he would be able to represent the facts much better and in your sense, and stop the rumours that would spawn because of insufficient information.

At a company where I worked, we installed closed circuit television to inform all our people when a plant or process upset had taken place, and the facts known till then. Before we had the CCTV, we used to hand out to the people at the gates when they left handbills containing the latest information. Some of the mishaps, such as gas releases or explosions, or even the warning signals sounded for some other minor mishaps, are noticeable outside, and the gossip factory starts painting catastrophic scenarios. In such cases our people, who were informed by us, could answer the questions and stop the damage. Others in the town expected the person working in the company to tell them what really happened, and whether it was dangerous for them. With our information system, our people were prepared for questions and could help spread the facts we had provided them.

At the top of safety committees is the Central Safety Committee, which may also be called the Safety Steering Committee. Let us now discuss when a Central Safety Committee is effective. First and foremost, a central safety committee is effective only when it is backed by top management's commitment. Remembering the three Is of commitment, we know that this means: (1) top management's interest, (2) its involvement and (3) its investment. Lack of any of the three is the death knell for the effectiveness of the Central Safety Committee.

Properties of an *effective* Central Safety Committee are:

- It has the topmost manager, e.g. the works manager, as its Chairman.
- Other members of the management are departmental heads or plant managers.
- All departments, including nonproduction, especially human resources, are represented by their managers.
- A representative of the works council or the union is invited to become a member.
- The secretary of the committee is the safety professional.

- Meetings take place at least once a month.
- The meetings do not last more than an hour.
- Regular attendance by all members, minimum at 11 of the 12 meetings, is mandatory.
- Representation of members by others is not allowed.

The Central Safety Committee is so important for the safety efforts at a site that existing safety committees, if not on these lines, should be disbanded as soon as possible and new committees as outlined above installed. The meetings are organised and hosted by the safety department, which also puts up an agenda in collaboration with the chairman and distributes minutes of previous meeting.

Agenda items for the meetings of the Central Safety Committee could be:

- Review of the latest safety statistics presented by the safety department
- Description of accidents or near-misses by the responsible departmental head
- Discussion on the root causes of these accidents and countermeasures planned
- A short presentation on a safety-related topic, e.g. noise and its effects, by an inside or an outside expert
- Discussion on unusual issues discovered during safety visits
- Deciding on specific safety projects, e.g. safety campaigns, monitoring their progress and choosing a sponsor amongst the members

The sponsor has the job of putting a working group together to work on the project, and promoting the team's work by obtaining support for it financially and time-wise. At least 50% of the working group members must be workers. The sponsor reports the progress at the Central Safety Committee's meeting.

Similar safety committees have to be installed at the departmental or plant levels and at sectional levels. The principles remain the same as for the Central Safety Committee, however, their frequency could be higher, and their duration could be shorter. Thus a departmental safety committee could meet fortnightly for about 30 minutes and a sectional committee every week for about 15 minutes. As mentioned, the meetings should also be used for communication of points discussed and decided at the Central Safety Committee level by members of the Committee.

The main difference between the Central Safety Committee and other committees is that in the latter, the majority of members are workers, and the chairman is a worker or a supervisor elected by the committee. Departmental heads or plant managers are compulsory members, but their attendance once a month is sufficient.

Agenda items for these meetings are closer to the work in the plant and section, and could be:

- Seeking solutions for specific issues of concern
- Discussion of accidents and near-misses that happened
- Discussing potential safety accidents that were discovered
- Communication of items gathered at the Central Safety Committee meeting, including financial and production/sales information (once a month by the attending manager)

In companies that have excellent safety performance, *all* meetings at *all* levels, not only safety meetings, have as agenda point number one "safety." This is the way to keep the flame alive. Remember, safety will only be as important as the managers make it. Safety meetings should not be the only time one speaks of safety. You can weave it into any discussion. As managers, you should ask time and again questions on safety. Let the people know that you put great importance on it. Once you start doing it, it will spread to all other levels, and safety will become the "must word" in your factory or company.

We have spent decades putting safety on a shelf where it can remain collecting dust till, once in a while, we brush off the dust and look at it, hurriedly putting it back in case we get dirty. Safety professionals with the "exaggerated" importance they put on their job had to be tolerated. With this attitude, they were seen by others as Salvation Army singers who could sing their song, get a little applause and go away without changing anything.

You, my dear managers, have the power to change this. It does not cost you more than asking a few questions once in a while and never forgetting to ask them. Just that little attention from you will reduce the injuries or even deaths in your area of management. Is that asking for too much?

24
Lock Out – Tag Out

In a factory manufacturing plastic material, granules of polystyrene were stored in large silos that fed the conveyer system to the bag-filling machines. The silos were cylindrical with a diameter of about three metres (10 feet), and a cone-like bottom that ended in a screw conveyer. The bearing of the screw conveyer had overheated, melting the granules near it and jamming the rotation of the screw. The jam inside the silo had to be removed by climbing into the silo (which had

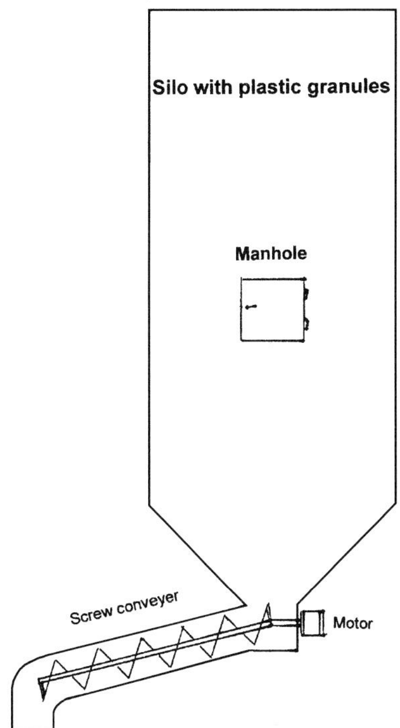

Figure 13 Silo

Managing Safety. Kishor Bhagwati

ISBN 3-527-31583-7

rungs inside), loosening the jammed granules and removing them by hand. The silo was shut off, and the maintenance department was asked to remove the jam.

The entrance to the inside of the silo was through a door of about 80 cm square in the cylindrical part. The door had a microswitch that hindered the machine being operated when it was open. The maintenance worker, who worked on the day shift, brought the main switch into the off position, climbed into the silo, went down the rungs to the bottom and started removing the jammed granules. Just as he was nearing the end of his work, the morning shift was replaced by the afternoon shift, and the new shift operator responsible for the area began on his normal round of the plant, saw the open silo door, closed it and turned on the power to the machine. The freed screw conveyer started rotating again, catching the man in its screw. The result is too gruesome to describe.

What had happened? The maintenance man relied on the microswitch and the main switch, which he had switched off, to stop any movement of the screw conveyer. He had not *locked* the main switch in the off position. About 40% of fatal accidents in the industry happen in this way! In spite of the fact that nearly every industrial switch sold nowadays has provision for putting a padlock on it, very few companies use them for the purpose for which they are there. I am always astonished at the ignorance of this locking system in factories that otherwise pride themselves on being safety conscious. This is especially notable in continental Europe, where the system is practically unknown.

The system is known as Lockout/Tagout, or in short LoTo. (In the US, there has been an OSHA Standard for this since October 1989[38]. In Europe, no such regulation or standard exists.) It very simply means that the maintenance or service person that has to open or work on a machine switches off the main switch, and locks it in the off position with his personal padlock, the key to which he keeps with himself. Thus as long as he is working, no one else can start the machine, either unknowingly or even accidentally (by inadvertently coming in contact with the start lever or the start button). He and he alone can re-energize the machine when he wants it. With a padlock costing a few euros, you can save a life!

38) OSHA Regulations (Standards – 29 CFR): The control of hazardous energy (Lockout/tagout). – 1910.147.

In a pharmaceutical company I visited, the rapidly rotating parts of tabletting and tablet packing machines were enclosed in Plexiglas housing with microswitch-protected doors to prevent anybody from putting their hand inside to remove a jam. The service mechanic, however, had to have energy to adjust and test the machine. For this, a key switch (like in your car) was built into the control panel. With the right key, the mechanic could switch on the electricity to do his testing, even if the doors were open. Without the key, one could not start the machine with open doors. Very clever and very comforting, were it not for the fact that in each and every machine in this plant, the key was always sticking in the key hole, irrespective of the fact whether the machine was being serviced or not. And I saw two lady operators turning the key to run the machine with doors open to remove with their hands a jam that had stopped the machine.

Upon my asking the mechanic, I got the following explanation: Each key was different, and it would be very impracticable having all these keys in the central workshop, from where the mechanic had to take it to service the machine. And if, by chance, he forgot to return the key to its place, his colleague in the next shift would not be able to carry out the next adjustment or repair, and the production would have to be stopped. The way to avoid this would be to have spare keys locked in a key box in the supervisor's office, hoping that the one person who has the key to the key box is present when he is wanted. A freely available spare key would completely destroy the purpose of the lockout system. I was assured that there was no reason to worry, because all the ladies working on the machines knew that they had strict instructions not to touch the key, contrary to what I had just seen a few minutes ago!

Some companies believe in entries in the plant log book to avoid such hazardous situations for service people. The service or maintenance person enters into the log book where he would be working, and how long he will be at it. He does not have to lock the switch in the "Off" position with his personal lock. All other workers are expected to read the log entry and not activate the switch for the area mentioned. This also assures the transfer of information at shift changes. Trust in such systems is the biggest hazard one can have. Those who introduced the log system had the feeling that they had now done everything to see that nobody gets hurt. And still accidents happen. Blame is put on the people and their forgetfulness or carelessness. How much simpler and foolproof is the method with personal padlocks! That the person who attaches the padlock also writes it down in the log book goes without saying.

If more persons are working on a machine or a process, there are attachments (hasps) that allow multiple padlocks to be brought on to the same switch. Each person working there puts his personal padlock on the hasp, and only when the last of them removes his padlock the machine can be re-energised.

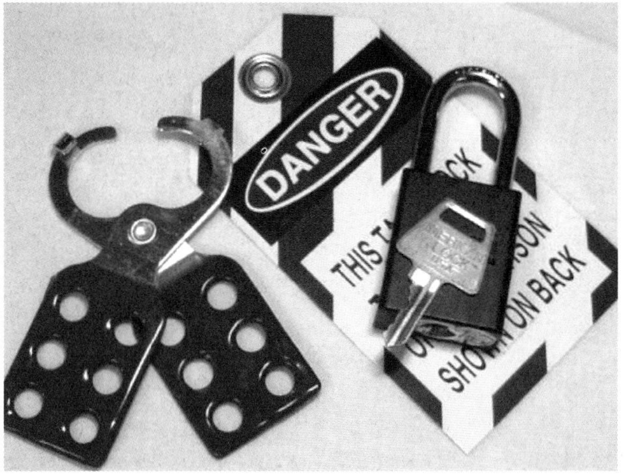

Figure 14 Locks

A similar system is also used for arresting flow of liquids where there should be no flow, by having a chain and a padlock system on the valve handle.

The plant operator has to know who locked the system. For this, the service or maintenance person who locks out the energy or the flow puts a tag at the switch or the valve that identifies him, when he stopped the current or the flow and when it will be released again. If a hasp is used, all whose locks are on it must identify themselves separately. Tags are also used when padlocks are not used. They bear warnings such as: “Do not start” or “Do not operate” or “Do not open the valve”.

Introduce this excellent and simple system at your site. You will be preventing many accidents, near-misses, and especially fatalities with this. Get your safety department to put up an internal regulation for this system and make it a point to watch out for such devices on your safety visits.

25
Communications

On a foggy evening of 27 March 1977, two Boeing 747s got ready for their take-off from Los Rodeos airport in the island of Tenerife, one of the seven Canary Islands off the west coast of Africa and belonging to Spain[39]. One plane was of the Pan American airline (Pan Am), named *Clipper Victor*, and the other of the Royal Dutch Airline KLM[40], named *Rijn* (river Rhine). Pan Am came from Los Angeles over New York and KLM direct from Amsterdam. Both were supposed to land at the major Las Palmas airport on the island Gran Canaria. However, because of a bomb threat at Las Palmas airport, the planes, together with three other large planes and several other planes, were redirected to the regional airport Los Rodeos on the neighbouring island of Tenerife.

This regional airport had only one runway and one parallel taxiway, with side lanes connecting the two (see Fig. 15). The taxi lane was full with airplanes that had been redirected and the Pan Am plane, which was still on the runway, was asked to wait in the side lane C-3. Meanwhile, the KLM plane, which had been given clearance for the route it had to take immediately after take-off, proceeded to the end of the runway in the starting position. The Pan Am machine was moving on the runway in the opposite direction, nearing the turn-off point into the side lane C-3. The pilot found the angle too acute to turn the huge aircraft and proceeded to C-4, which had a less acute angle.

39) The name of the archipelago of seven islands is derived from the Latin term Insularia Canaria meaning Island of the Dogs, a name applied originally only to the island of Gran Canaria. It is thought that the dense population of an endemic species of large and fierce dogs was the characteristic that most struck the few ancient Romans who established contact with the islands by sea. Today's coat of arms of Comunidad Autónoma de Canarias shows a shield supported on each side by a dog, the shield showing the seven mountainous islands. The bird canary is a native of the islands and is named after them – not the other way around.

40) KLM = Koninklijke Luchtvaart Maatschappij, literally Royal Aviation Company, since 2004 part of Air France.

Managing Safety. Kishor Bhagwati

ISBN 3-527-31583-7

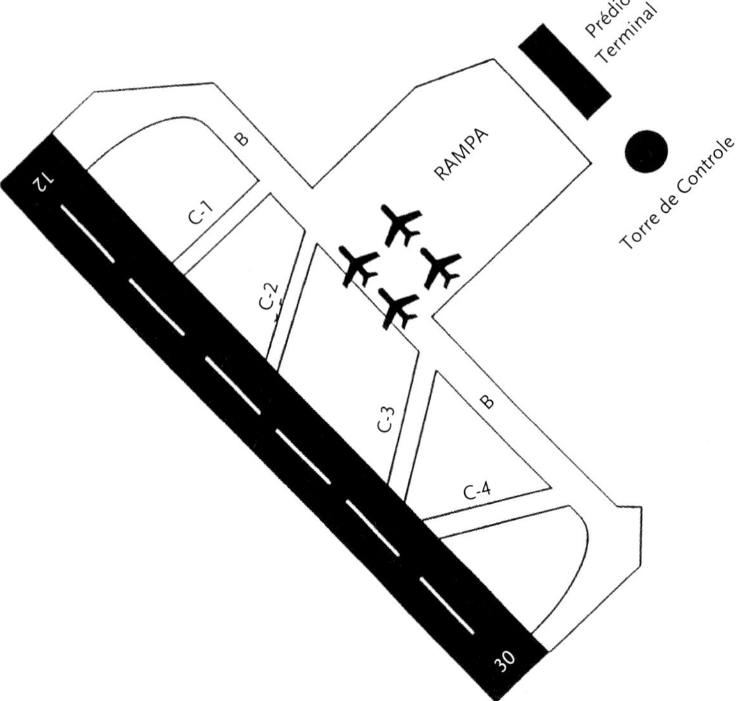

Figure 15 Tenerife Airport

Meanwhile, KLM radioed to the tower that they were "at take-off". The tower asked Pan Am whether they had left the runway, to which Pan Am replied in the negative. The tower, therefore, asked KLM to wait, expecting that they had not yet started the take-off. But the first officer of KLM, whose command of the English language was not very good, had said, as they say in Dutch, that they were "at take-off" meaning the English term "taking off". Due to a shrill noise in the wireless communication (both planes were on the same frequency), neither plane had heard the other's message. KLM started its run, the pilot saw the lights of the Pan Am too late due to the fog, tried to pull his machine up, but could not avoid its tail hitting the fuselage of the Pan Am plane. Both planes were engulfed in fire and 583 persons lost their lives. This was the worst accident in the history of aviation.

Figure 16 747 Jets

Figure 17 Plane wreckage

There were other factors that played a part in this disaster, but the main problem was found to be the communication between the tower and the two planes. Undefined word usage, not asking for confirmation of what was heard, and assuming that whatever one said was received, understood and okayed by the other, were at the root of the catastrophe.

Managers often live under this delusion. They think that whatever they have said has reached its destination and has been understood and accepted by the recipients. They also believe that their oral communications via the proper channels, i.e. through the chain of their subordinates, will carry their thoughts to the lowest level with high fidelity. The truth, however, is that oral communication and messages, percolating through the hierarchical levels, hardly carry the spirit and the impact of the message, and seldom also the content. Each level filters the content till what reaches the worker is often something different from what the top management had in mind. In the other direction it is still worse. Things that workers would like the top management to know seldom ever reach that level. They are generally strangled half-way up by the intermediaries. The following diagram illustrates this.

TOP-DOWN		DOWN - UP
⇩ ⇩ ⇩ ⇩ ⇩	Top Management	0
⇩ ⇩ ⇩ ⇩	First Level Mgrs	↑
⇩ ⇩ ⇩	Second Level Mgrs	⇧ ⇧
⇩ ⇩	Supervisors	⇧ ⇧ ⇧ ⇧
↓	Workers	⇧ ⇧ ⇧ ⇧ ⇧

Figure 18 The Fate of Communications: Top-Down and Down-Up

There is nothing unnatural about this. As a matter of fact, it is the most natural thing. Do you remember the game we used to play as children where all set around in a circle, and one whispered something in the ear of the child next to him? He then whispered what he heard and understood to the next one, etc.? What came out at the end was entirely different from what the first one had said. A communication down or up the hierarchy chain suffers a similar fate. It is filtered at every stage, reducing its message content.

What is the way out of this? Very simple: Top managers should bypass all hierarchical levels by talking directly to the worker. This is what we have called a safety visit! Safety visits bring the management in direct contact with the workers on a one-to-one basis, bypassing the intermediate hierarchical levels that filter and distort oral messages.

The main purpose of a safety visit is to establish a free exchange of ideas between the top and the bottom levels, bypassing the "proper channels". As we have seen in the chapters on safety visits, one does a safety visit more to listen than to talk. Thus the greater part of the conversation during a safety visit is down-up, rather than up-down. You will be surprised at things you get to hear. You also have the opportunity of checking whether your communication from the top has reached its target by asking the workers whether they had any comments to your latest oral message.

With that we come to written safety messages from the management. Your workers do not report to the safety department, they report to you. Messages from the safety department do not carry the weight your messages do. It is, therefore, your duty to regularly publish safety messages. You can ask the help of your safety department for ideas and texts, but the message should come from you and your name should be at the bottom of the page. The messages should be short, simple and easily understood by your readers.

Communication is not what you have said or written, but what your listeners or readers have understood and accepted.

The language used by lawyers is not meant to be understood by the laymen. The language used by government servants is still worse, although it is meant to be understood by laymen with whom they correspond. Worst is, however, the language used by technicians when they want to sound official. Descriptions and operating instructions for machines, especially computers that we all (have to) use, are full of jargon and googly sentences that make working with them almost like reading a strange language. They feel that they have rendered what they want to say extremely important by making it impossible for non-specialists to understand. Jamming your messages with official-sounding words and making sentences go over four lines has the same effect; they are not understood by those whom you want to address.

This is why you should speak in your listeners' language and write in your readers' words. You should remember that the majority of your shop-floor workforce is not avid readers of books. They seldom have more than a dozen books at home. Their main reading consists of tabloids of the popular press and not high-brow newspapers. You are not communicating with them if you write in a learned, semilegal language. They do not understand it, and still less accept it.

I once compared the same news in two different German papers. One was the tabloid "Bild" and the other the prestigious "Frankfurter Allgemeine Zeitung" (FAZ). No sentence in Bild had more than 8 to 9 words, and none of the words had more than 4 to 5 syllables. FAZ, on the other hand, had long sentences with 30 to 40 words containing learned multisyllable words.

Or compare the following two examples from British newspapers reporting on robbery by a gang that took nearly £50m from a cash depot in February 2006:

The Guardian: "The robbers who carried out the biggest cash raid in British history were the targets of the largest-ever reward last night as police attempted to entice criminals on the fringe of the gang to inform on its main members." (40 words)

Daily Express: "Police hope that the massive reward will encourage members of the criminal fraternity to turn informer." (16 words)

If you want to communicate with readers of Bild or Daily Express, which language should you use? That of the FAZ or The Guardian, or that of Bild or Daily Express?

Martin Luther (born in 1483 in Eisleben, Germany) was not, as is normally assumed, the first one to translate the Bible into German. His translation, published 1534, was the 19th translation of the Bible into German, the first being a translation by Johannes Mentelin of Strasbourg in 1466. What differentiated Luther's translation from other translations was the language it was written in. Whereas previous translators translated the official Latin Bible word for word, so that it could be understood only if one knew Latin well, Luther used the original Greek and Hebrew sources, and translated them into the language in which the people of the time spoke[41]. Suddenly, ordinary people could understand what was written in the Holy Book, his Bible

41) The first English translation on similar lines was the Geneva Bible of 1560, written by protestant scholars who had fled England to escape from Mary Tudor and had settled down in Geneva. The King James Version, commissioned by King James I of England in 1604, was first published in 1611.

became a best seller, and Christianity went through its greatest upheaval, splitting the Christian world in two. This is the impact of simple words.

Sir Winston Churchill was also known for his use of simple English. He never used the so-called "King's English". His famous short sentence after the Royal Air Force routed the German Luftwaffe during the Battle of Britain has gone down in history as the fullest (albeit shortest) praise ever given: "Never in the field of human conflict has so much been owed by so many to so few." Wordiness is good for disguising clarity, and is usually used for such purposes. Sir Winston said, "All great things are simple, and many can be expressed in single words: freedom, justice, honour, duty, mercy, hope."

Avoid lengthy, high-sounding, legal-looking texts, and you will be able to get your message through to the workers. All things, from messages by the management including company policies to safety instructions and even operating instructions, are victim to this lemmings-like drive to verbosity, mainly because they are written by those who do not have to follow them. The solution to this is: ask your workers to write these things in their own language. For example form a team to write, say, the operating instructions for a machine. The team should consist of workers who operate this machine, and a few persons from the engineering and maintenance departments. The actual writing should be done by the workers, the technicians just checking whether it is technically correct. You can even do this with the Company Safety Policy, and watch the simple and understandable sentences, that express the same thing, come out. Instead of sentences like: "All employees in leadership positions are responsible for deploying and actively supporting the Company's overall Safety Program that includes their responsibility for monitoring and enforcing the use of safe working practices and safety rules.", you will have sentences like: "Managers are responsible for the safety of their subordinates."

I found the following nice example of an official-sounding puffed-up text in my scrap-book:
"Ladies and gentlemen, we will momentarily be landing in the Atlanta area. We're sorry that, due to operational problems, we'll be experiencing a late arrival. At this time please extinguish all smoking materials[42]. We will be deplaning by the forward exit. If Atlanta is not in

42) In those times, smoking was still allowed in airplanes.

your travel plans this evening, and you are catching a connecting flight to another city, you may suffer an involuntary boarding denial. If so, one of our customer personnel will be happy to take care of you.....blah.....blah blah. ...Thank you for flying Tin Can Airlines."

Translation: "In a moment, we'll be landing in Atlanta. Sorry we're late: the aircraft broke down. Please put out your cigarettes. You will get out at the front. If you have missed your connecting flight, one of our staff will help you."

26
Other Managerial Tools

We are now approaching the end of this book. Apart from the tools described in the preceding chapters, there are a few things that round up the safety effort and that a manager should be basically aware of. I shall go into them briefly, highlighting what you should look out for and where you, as a manager, can give a hand to see that things are done in the right way.

The Permit-to-Work System

I am sure you have a permit-to-work system at your site. For any repair or maintenance work, or making changes in the plant, a permit has to be issued by the plant where the work is to be carried out. The permit certifies to the outside worker that the plant has taken care to isolate – energetically and product-wise – the part of the machine or plant where the work has to be done. Normally, the permit is required only by contractor workers, and not by company's own maintenance workers.

A permit must be issued and required by *all* workers, whether own, or from outside. Many accidents have happened because the plant did not know when the central maintenance workshop man was carrying out a repair, and started the process, spraying the man, who had opened a pipeline for some work, with corrosive chemicals.

A permit lists all precautionary measures that have to be taken before the service or maintenance work in a plant is started. The precautionary measures are listed (usually in very small print), and whether they are needed or not has to be ticked in the box provided against the measure.

Managing Safety. Kishor Bhagwati

ISBN 3-527-31583-7

Check whether there are two columns for the boxes, one for what is required, and the other for measures that are not required or not applicable. If there is only one column for boxes, then one cannot know whether the measure is required but forgotten, or not required. For each measure one of the columns must have a tick.

Check whether the lockout/tagout system, described in chapter 23, is listed as a compulsory measure on the permit.

The validity of the permit is mentioned on the permit. For longer works, permits are often issued for several days. We read about the fatality in the silo that happened at shift change. This is quite frequent, the time of shift change being very critical. If a contractor has to work for several days, e.g. when carrying out some civil work, then that area should be cordoned off and considered exterritorial, and put under the supervision of the contractor.

A permit to work should be valid only for one rotating shift. It can be renewed once for the following shift, but not longer. The permit should be displayed visibly near the place where the work is going on. I introduced in my plant permits printed in three different colours, one colour for each shift, so that one can recognise from far whether it was valid for the running shift, or not.

The permit is to be authorised by the plant manager or his designated supervisor. I have quite often seen how this is done. The outside worker comes with the permit form filled out by him into the office of the manager or the supervisor who glances at it, asks the man seeking the permission whether he is going to follow all the instructions on the permit, and signs it sitting in his office.

The person who signs the permit must put his signature under the following statement at the bottom of the permit: "I have personally inspected the place where the work will be carried out, checked whether all listed safety measures are implemented, and found them to be in order."

The permit has to be signed both by the plant representative and the worker (or the work-team leader).

When doing your safety visits, do not fail to look at the permits and ask the worker whether he has understood all that is on the permit, and agrees with the safety measures prescribed.

Operating Procedures

When making your rounds, ask the worker to show you his operating instructions (they should be available immediately at his place of work, and not in the foreman's cabin) and look when the last revision or updating had taken place. Do not be shocked when you see that the last revision was 10 to 12 years ago. I have seen operating instructions that are still older, and do not reflect in any way what the process or the plant looks like at present. They are like distribution lists in office mails, which go on containing names of people who have already left the company, or even of those who are dead.

Procedures and drawings must be updated every two years at the maximum. If an update is not necessary, they should at least be gone through again to see whether everything is as is written. The date of revision and the next date of revision should be written on every page of the procedure.

As Trevor Kletz writes: "Procedures, including testing and maintenance procedures, are subject to a form of corrosion more rapid than that which affects steelwork and can vanish without trace in a few months once managers lose interest."[43] Accidents like Bhopal happened because of this problem.

Contractors

In the first exercise in courses I give, participants have to write down which issues in their factory they are worried about. Invariably, in the top three of these, is the problem with contractors. With the extensive outsourcing going on, there are factories where nearly 50% of the workforce is from contractor companies.

43) Trevor Kletz, loc.cit.

We do not want to go in depth into contractor management here, but as far as safety is concerned, we can achieve a lot by incorporating a few things in the process of selecting and managing contractors.

1. Look at the safety record of the contractor. It should play as important a role in his selection as his price.[44)]
2. Demand from your contractor an SCC certification. SCC means Safety Certificate for Contractors. In 1989, Dutch petrochemical companies in the Rotterdam Europoort area developed a standardised single checklist to replace the many different checks by several companies to assess the occupational health and safety (OHS) performance of their contractors[45)]. A few years later they introduced procedures and guidelines, out of which was developed an independent certification system with 10 topics, 54 questions and 140 criteria. Many countries in Europe have adopted this system, which is excellent for judging the safety management qualities of a contractor, and gives the customers, i.e. you, the assurance that the safety efforts you have put in your company are not undermined by the contractors.
3. The contractor is responsible for instructing his workers on the safety standards valid in your company. He is also responsible and accountable for seeing to it that they are followed. This has to be mentioned in the contract.
4. The contract should include the clause that if contractor personnel disregard any of the safety regulations of your company, the contract will be terminated with immediate effect. This should not only be on paper, but should be exercised if the case arises, to show that you mean it.
5. At the time of signing the contract, your purchaser should go through it clause by clause with the contractor to assure that it is understood by him. Contractors will sign anything to get the contract, and it is not our purpose to show him a signature if an accident happens, but to avoid the accident in the first place.

44) I remember a cartoon showing two astronauts sitting in their seats in a rocket about to take off, and one astronaut saying to the other: "Do you realise that we are going into space with a rocket built of parts supplied by the lowest bidders?"

45) In the Netherlands it was initially called VCA = Veiligheid Certificaat voor Aannemers.

6. Finally, who is also responsible for the safety of a contractor worker? You – and your people! As long as they are in your plants, they are in your area of responsibility, and you have to see to it that they do not get hurt.

Rewards and Incentives

Many top managers believe that safety competition among plants motivates people to do more. They offer rewards and certificates for the best safety performance, praising the plants and their managers in company bulletins and showering them with praise. Unfortunately though, a factory is not a playground, and safety performance is not a race or a match. Putting a reward as the carrot dangling in front of a plant manager distorts his view of safety as something that is not normal.

A worker is expected to work in a productive manner, produce high quality and work safely. These three things are equivalent. Why reward something that is *normal*? As Thomas Krause[46] puts it: "No sane manager would try to improve production or quality with such 'incentives', so why do it in safety? Because until recently safety performance was thought to be due to a 'special cause' unlike other aspects of industry or business." Safety was considered as something on which one spent a little time when everything else was done and then there was some time left. It was like doing overtime, and overtime has to be rewarded!

Rewarding safety performance has one other negative aspect. The basis for the reward is usually the accident frequency rate (number of accidents per 100 or 1000 employees)[47]. The plant or the department, or on corporate scale the factory, with the lowest accident rate gets a reward. In a factory I know there were two plants, plants A and B, run by two very active and committed plant managers. They both worked a lot on improving safety in their plant, and were neck-to-neck in their accident figures, which were quite low. Both were looking forward to getting the reward this year, and eyed surreptitiously at the other plant's performance every day.

46) Thomas R Krause, loc. cit.
47) See Appendix 1.

Then, on 30 December, plant A had an accident. Plant B, of course, got the reward and was praised by the works manager at the annual meeting. Can you guess how the manager of plant A felt? Do you think he was more motivated now? Did he have any incentive left? Remember: For every winner, there are several losers. We do not want to make our people feel that they are losers. Before I took over the job of SHE Director for the European Division of a US company, there was an annual rite of declaring first, second and third winners among the 28 plants in my division, based on their accident rate for the year. The winners were mostly plants in the UK, Germany or France. On my first visit to the Spanish plant, the works manager and I started discussing safety, and he said it was no use doing anything in this matter, because he was at the tail end of the list anyway, and he never expected to win a reward. He was dejected because every year, at the presentation of the safety statistics and award winners at the January board meeting, everybody looked down upon him and made him feel ashamed. One would have expected him to be driven with the desire to win a medal once, but it did not work out that way. This is how the human being is!

Celebration is something else. If the *site* had a good safety result, all at the site should celebrate. This could be receiving small presents, such as a steam iron for the home or a battery charger for the car, etc., but *never* money! It should be a memento of safety, and not a payment for it. Never have a safety competition amongst plants at a site. It never motivates.

Some sites have put up big boards at their entrance showing the number of accident-free days the site had. This is a good idea, but it gets spoiled if one also breaks down the site into various plants, and shows the numbers for each plant individually. With this you are entering into the conflict zone with negative effects.

The driving force behind safety must be the inner will, and not the monetary gain. The wish should be to be safe and protect oneself from injuries, not to get a reward. And it should be considered the *normal way of working*. Professionals just do *not* work in an unsafe manner. One neither expects, nor gets a reward for working safely. Let your company make this clear at the time of employing the worker. Make safe working a condition for employment.

27
How to Proceed

Safety is a vast subject. There are many aspects and much knowledge above and beyond what we have discussed in this book. But all that is for your safety professional. For the purpose of *managing* safety, things we have gone through until now are sufficient. This book will not make a safety expert of you, but will open your eyes and provide you with the tools you will need to bring order in safety efforts of your company and give you guidance on how you can optimise your safety system to produce the most desirable results.

I give below two "To Do" lists, depending on the management level you belong to. The first is for the top management, usually the works (site) manager, and the second is for the middle management in production, like plant and maintenance managers. If the whole corporation wants to change its style of safety management, even the CEO can adopt this list with gain.

To Do List for the Top Management

- Install a Safety Steering Committee under your chairmanship. For its structure and agenda points see Chapter 23 on Safety Committees.
- Get yourself and all your managers trained in correct safety visiting techniques (theory and practice) and in the principles of accident investigation.
- Rewrite the job description of the safety professional as described in Appendix 1.
- Make the safety professional report directly to you.
- Announce his job description to all employees at the site, stressing his function as a consultant and not a doer, and his unavailability as a dogsbody for safety matters.

Managing Safety. Kishor Bhagwati

ISBN 3-527-31583-7

- Clarify to all managers what their safety duties are, and add these duties in their employment contract.
- Include safety management as an important criterion in managers' annual performance evaluation, with a weighting of 20% to 25%.
- Get the HR department to include safe working as a condition of employment in all employment contracts with workers.
- Prepare your safety visit schedule in cooperation with the Safety Department and publish it 6 months in advance.
- Have the safety professional present at the management meeting once a quarter all managers' degree of fulfilment of their safety visit quotas, and the quality of visits.
- Demand that all accidents are known by you within 24 hours.
- See to it that all near-misses are recorded and investigated.
- Get copied on all accident-investigation reports and check whether they guarantee a nonrecurrence of the same or a similar accident; if not, send them back.
- Visit an injured person who has returned to work during your safety visit.
- In the case of serious accidents, send flowers and a get-well card, or, in very serious cases, visit the victim in the hospital.
- Take every opportunity of talking about safety with your employees.
- Get signs printed and exhibited everywhere with the management message in Chapter 5.
- Get the lockout/tagout system introduced via the safety department.
- Get the permit-to-work system revised by the safety department on lines described in the previous chapter.
- Replace the recognition and reward system for individual plants by celebration for all.
- Stop displaying a plant-wise breakdown of accident-free days at the site entrance.

To Do List for the Middle Management in Production

- Install a departmental or plant Safety Committee comprising 50% workers.
- Get a worker elected as the chairman for a two-year period.

- Attend at least 90% of Safety Committee meetings.
- Prepare your safety-visit schedule in cooperation with the safety department and publish it three months in advance.
- Scan safety-visit reports of your subordinates to see whether more unsafe acts are found than unsafe conditions.
- Demand immediate *personal* reporting of accidents.
- Be at the site of a serious accident as soon as possible.
- Participate in accident investigation of serious accidents as an observer.
- Get your foremen and supervisors trained in nonfault-finding and nonpunitive accident-investigation techniques.
- In case of a serious accident, visit the injured in hospital and comfort his family.
- Get yourself and your supervisors trained in brainstorming techniques.
- Start all other meetings with a safety message.

To Do List for Nonproduction Management (HR, IT, Finance, etc.)

- Get yourself trained in carrying out safety visits.
- Get your senior staff trained in basic safety principles.
- Prepare your safety visit schedule in cooperation with the safety department and publish it six months in advance.
- Safety should be agenda point number 1 in all departmental meetings.

Once all the above points are implemented, your company, site or factory is on the right track to achieving excellence in safety. I promise you very satisfying times and the type of inner happiness a doctor must feel when he has cured a patient of a consuming disease.

What comes after this? Once all the manager-driven safety activities are functioning properly and you have reached a plateau in your accident frequency curve, it is time to take the next step, and that step is behaviour-based safety (BBS). Managers can carry out their safety visits and discover unsafe acts, but they are not always there. But who is always there, is the colleague. BBS trains workers in observing each other and giving each other feedback on their observation. It is quite a complex programme, as the matter is quite sensitive and could backfire if not carried out by an expert in the field. With BBS, you can reach still greater heights, culminating in zero accidents.

It is time to start. Do not delay, because a delay may mean an accident that was not stopped in time. I wish you the best of luck. And remember:

Either it starts at the very top, or it doesn't get anywhere.

Appendix 1:
Understanding the Numbers in Safety Statistics

Let us take two factories whose safety statistics we want to compare. Factory Alpha had 20 accidents last year, and factory Beta 30 accidents. Does that make Alpha's safety performance better than Beta's?

One cannot compare absolute numbers, only relative numbers. If Alpha has 200 employees and Beta 400, then the ratio of accidents to employees becomes 20/200 = 0.1 or 10% for Alpha, and 30/400 = 0.075 or 7.5% for Beta. Thus Beta's performance is better than Alpha's.

To be able to compare various factories, sites and sectors with each other, one has to base the number of accidents on defined reference values, such as the number of employees or their hours of work. "Employees" means all persons at a site on the company's payroll, but does not include contractors. This is used by some companies to reduce their number of accidents by employing contractors to do accident-prone jobs. A sudden large step-like drop in a site's accidents is a sure sign of such a trick. Companies who are in the first league of safety performance and have the right safety attitude, i.e. companies who believe that the safety of a contractor is also their responsibility, include contractor accidents also.

Reference points vary from country to country. In the US, one uses as reference the one prescribed by the OSHA[48], which is 200 000 man-hours per year. This number is based on the assumption that a worker works 40 hours a week for 50 weeks a year, which equals 2000 hours per year[49]. A standard factory with 100 workers will thus log 200 000 hours. All accident statistics, therefore, are based on a standardised factory size of 100 employees.

48) Recording and Reporting Occupational Injuries and Illness, OSHA, Department of Labor, 2001.

49) West European companies would rejoice if they could bring their workers to work 2000 hours per annum. With all those vacations and holidays, they average about 1600 to 1700 hours per year!

Managing Safety. Kishor Bhagwati

ISBN 3-527-31583-7

The number of accidents based on a reference point is called the frequency rate or the incidence rate. The accident frequency rate for a factory with, say, 300 employees working 600 000 hours and having 50 accidents in a year will be:

(200 000 / 600 000) × 50 = 16.7 lost time (LT) frequency rate (LTFR) or
lost time incidence rate (LTI or LTIR)

In other (and simpler) words, this means that about 17 % of your workers had an accident in the past 12 months.

Other reference points, especially in Europe, are based on 1000 employees, or a million man-hours. To approximately compare these figures with OSHA type figures, just divide the thousand employees number by 10, and the million man-hour figure by 5.

In the OSHA system, if an employee is not able to work for one workday *following* the day of the accident, the accident is to be counted ("recordable" accident). First-aid cases are counted only under certain conditions. In Europe the absenteeism has to be for three days. OSHA does not count accidents on the way to and from the normal place of work, whereas in Europe, these are considered work-related accidents and counted. Internationally operating European companies have adapted the OSHA system for internal use with some changes (e.g. for medical treatment and first-aid cases), but report to their authorities as per local regulations.

Thus, when somebody shows you frequency rate numbers, ask what is the reference point being used and whether accidents on journeys to and from the place of work, and contractor accidents are included.

Frequency rates are statistical numbers, and underlie fluctuations. A drop in the frequency rate from, say, 5.0 to 4.5 in one year is within the range of normal variations and no reason for celebrating. Unless there is drop of over 50%, and this is sustained for at least three years, you cannot claim to have improved your safety performance substantially.

Appendix 2: Job Description of a Safety Professional

Treat your safety professional as your Minister of Safety. Your minister who advises you and organises for you. See to it that he does not become the person on whom managers unload their responsibility for safety. Now that you know how good superior safety is for your business, let his knowledge and professionalism help *you* achieve *your* goal of smooth, uninterrupted production.

Below are the main points you should include in his job description:

- He reports directly to the top management without an intermediary.
- He acts as a safety consultant for all managers at site.
- He is not responsible for the safety performance of the site.
- He does not carry out accident investigations, but assists if required, especially during investigations of serious accidents and fatalities.
- He coordinates site-wide safety activities.
- He coordinates safety visit schedules of managers and evaluates their visit reports.
- He presents the fulfilment of safety visit quotas of managers at management meeting.
- He carries out safety inspections at his own discretion.
- He collects safety statistics and evaluates them to discover trends and gaps.
- He submits site safety statistics to the authorities and interacts with them on all safety related matters.
- He arranges training of workers by inside or outside experts at the request of plant managers; administration of the training sessions is done by the plant.
- He takes part in project meetings to examine safety aspects of designs.

Managing Safety. Kishor Bhagwati

ISBN 3-527-31583-7

- He advises on safety equipment for special hazardous activities.
- His approval is needed on all new purchases of safety materials.
- All new equipment and machines have to be reviewed by him before purchase, and he can prescribe changes or additions to make them safer.
- He carries out safety inspections on long-term contractor activities.
- He manages, coordinates and takes part in outside audits, e.g. corporate audits and by certifying institutions.
- He writes the permit-to-work procedure and its forms, and inspects their use periodically.
- He introduces the lockout/tagout system, and inspects its use periodically.
- He manages the company fire brigade or fire specialist.
- He indoctrinates all new employees in safety principles of the site and applicable safety rules.
- He has an up-to-date library of safety-related literature and regulations.
- He represents the site or company at safety-related work groups of professional societies and industry federations.

His job description is to be made known to all line managers.

Appendix 3:
Safety visit Reports

On the following pages you have two different safety-visit reports. The first is a report by Mr Saras, the plant manager of the Assembly Plant No. 3, and the second a report by Mr Malfoy, plant manager of the Can Manufacturing Plant. One of these is a report as it should be, and the other as it should not be.

Study both reports and decide which of these reports is as it should be, and what differentiates it from the report as it should not be. As a manager you may not have time to read all the details in the reports you receive, but you must know how a good report should look. You only have to read the reports of those who report to you directly, thus it is not too much time we are talking of. But you should be able to discover at a glance whether the report is good or not, so that you can use the next opportunity of talking to the person who submitted a not-so-good report.

The following check-list provides you the tool to discover the quality of a safety visit report at a glance:

1. Check whether a worker is included in the team.
2. Check whether there is not too much delay between the visit date and the date the report was received in your office.
3. Check the number of persons the visitor talked with. Four persons in half an hour is a good average.
4. Check whether there are any positive remarks. A visit should also highlight safe actions and conditions, especially improvements from the last visit.
5. Check the ratio of acts to conditions observed. A ratio below 1.0 is not that good. A ratio below 0.5 should signal a yellow light, and one near zero, a red light.
6. Check whether the follow-up item lists the name of the person who has to take care of it, and a deadline that is a date.

Managing Safety. Kishor Bhagwati

ISBN 3-527-31583-7

Make a note of one or two points you may want to look at during your next safety visit to the section.

An additional advantage you will have is that you will get to know the managerial qualities of your subordinates much better, and can follow their progress over the year. This will help you in evaluating his performance at the end of the year.

Safety Visit Report	Plant: Assembly 3	Section: Workshop
	↓ Name	↓ Position
Safety visit by	B. Saras	Plant Manager A3
Accompanied by	1. S. Dreher 2. M. Scott 3.	Turner Workshop supervisor
Visit date	08 March 2006	Duration: 35 minutes
	No. of persons talked with	5
	No. of persons working in the visited area	12

Findings S = Safe U = Unsafe	Act	Condition
1 Cleanliness and orderliness are good in the area and on work-benches		S
2 Sheet metal corner on work-table protruding beyond edge could hurt someone		U
3 Trainee holding the work piece with gloves while drilling on the drill bench	U	
4 Welder leaving the welding torch burning on the ground while fetching welding rods from the cupboard. Turned off the torch.	U	
5 Third rung from bottom of the ladder in store rusty and loose		U
6 Worker overreaching beyond the ladder to the right to get a spare part from a high shelf	U	
7 Worker found leaving a designated smoking area with a burning cigarette	U	
8 All other persons observed were wearing their PPE correctly as required for the job	S	
Totals	5	3
Acts observed / Conditions observed =	**1.7**	

Follow-Up of items listed above	Person	Deadline
2. Sheet pushed back. Agenda point next safety meeting	M. Scott	15 March 2006
3. Talked with trainee about the consequences for him	B. Saras	on the spot
4. Waited for welder to return, talked with him and asked him how to avoid the fire hazard in future	B. Saras	on the spot
5. Rung marked with red/white tape. To be repaired asap	S. Dreher	10 March 2006
6. Asked the worker to climb down and talked with him about hazards of falling and the consequences	B. Saras	on the spot
7. Explained to the worker why the non-smoking rule was made. Mark-off smoking area on ground with yellow tape.	M. Scott	08 March 2006

CC: Dept. Manager, Plant Safety, Central Safety Department, Plant Notice Board

Safety Visit Report	Plant: Can Manufacture	Section: Packaging	
	↓ Name	↓ Position	
Safety visit by	R. Malfoy	Plant Manager CM	
Accompanied by	1. K. Davidson 2. M. Duchamp 3. A. Bertollini	Section Head Packaging Foreman Safety Department	
Visit date	10 March 2006	Duration: 20 minutes	
No. of persons talked with		2	
No. of persons working in the visited area		36	

Findings S = Safe U = Unsafe	Act	Condition
1 Dispenser for ear plugs near the entrance empty		U
2 Worker seen driving his FLT too fast	U	
3 Empty pallets lying around in a disorderly manner		U
4 Cold water dripping onto machine from overhead pipe		U
5 Protective cover of can stamping machine removed		U
6 Used solvent rags in open buckets without lids		U
7 Housekeeping quite poor, especially in paint shop		U
8 ---------		
Totals	1	6
Acts observed / Conditions observed =	0.16	

Follow-Up of Items listed above	Person	Deadline
1. Dispenser to be replenished	Safety Dept	asap
4. Leak to be stopped	Maintenance	this month
5. Replace protective cover	Plant	April
6. Close all buckets	Plant	by tomorrow
CC: Dept. Manager, Central Safety Department, Plant Notice Board		

Accident Investigation Report		Plant: Product Flow	Section: Tank farm
Accident No. 321PF			
Date: 13 March 2006	Time: 14:50	Weekday: Monday	
Job of injured person	Maintenance mechanic		
Type of injury	Cut in left calf		
Treated where	Town hospital, as stitches were required		
Expected days of absence	3 days	Recordable? Yes	
Preliminary information distributed on: 13 March 2006		at: 15:45	

What happened

The mechanic tried to remove the sheet metal lid over a rectangular oil separation pit (1 m x 0.7 m) though which rainwater from the diked areas in the tank farm flows before meeting the main rainwater drainage pipe. The oil collected in the separator has to be removed from time to time by lifting the lid and suctioning off the oil into a container. The lid was jammed and the mechanic tried to pull the heavy lid with the grips provided on it, when it suddenly gave way and hit the mechanic in lower leg. Because of the weight of the lid and the force used to remove it, it resulted in an open wound which had to be treated in the town hospital.

Cause of accident

Although a hook is provided for this purpose, the mechanic opened the lid without using it because he had forgotten to take the hook with him, and his leg was thus nearer to the lid than it would have been if he had used the hook. He saw no necessity of fetching the hook because he had opened the lid quite often without the hook without any problem. A discussion with the mechanic revealed that it need not be necessary to fully remove the heavy lid for retrieving the oil, because a smaller opening would also suffice to insert the suction pipe to empty the collected oil. The mechanic suggested making a small (30 x 30 cm) hinged door with a grip in the lid for future use. His suggestion was accepted.

Countermeasures	Person	Deadline
1. Provide for hanging the hook next to the ditch in case the whole lid needs to be removed	H. Glossop	14. 03. 2006
2. Prepare work order for the workshop to make the hinged door in all five lids in the tank farm by 28 March 2006	M. Suleyman	17. 03. 2006
3. Check whether all the lids have been changed accordingly	M. Suleyman	29. 03. 2006
4. Discuss the case at the next safety meeting and stress the use of hooks to remove large lids. Ask for any other suggestions	P. Smith	20. 03. 2006

Investigation Team (person, position)

1. H. Glossop, Operator tank farm
2. P. Smith, Supervisor tank farm
3. M. Suleyman, Foreman tank farm
4. Victim after his return to work on 17. 03. 06

Start of investigation	End of investigation	This report date
13. 03. 06 and 17. 03. 06	17. 03. 06	18. 03. 06
Distribution: Works Manager, Plant Manager, Plant Safety Person, Central Safety Department, Notice Board		

Appendix 4: Accident investigation Reports

As in the case of safety visit reports, we have here two accident investigation reports. These two are reports of two different investigation teams that have investigated the *same* accident.

The first report is by the investigation team with Mr P Smith in it, and the second by the team with Mr Wooster in it. Both teams were faced with the same facts, but their approaches are different, and consequently also their results of cause finding and their solutions for preventing a recurrence.

Below are some hints that will help you determine the quality of the investigation. If you do not find the investigation thorough and proper, send it back immediately to the person in whose area the accident happened and ask him to repeat the investigation with the final aim of the report in mind, i.e. would the countermeasures suggested suffice to prevent a recurrence of the same or a similar accident in future?

Compare the accident-investigation reports by first studying them and then going through the following check-list:

1. Check the dates of the accident, that of the distribution of the preliminary information, the start of the investigation and the report date. These should be as close together as possible, usually not more than a few days to a week.
2. Check whether the box for "What happened" is clear enough and not too verbose or vague, and whether it contains any criticism of the injured or other workers.
3. Check whether the box for "Cause" puts blame on someone for having done something wrong. Blame is never to be accepted as a cause.
4. Check whether "carelessness" or "loss of attention" is given as a cause. These "causes" are unacceptable.

Managing Safety. Kishor Bhagwati

ISBN 3-527-31583-7

5. Check whether each countermeasure suggested has the name of a person associated with it, and a definite deadline, i.e. a date attached to it.
6. Check whether the investigation team included a colleague of the injured, and, if available, the victim himself.
7. Check whether the central safety department was included in the team, although the accident was not a serious one.

You can check all the above in approximately two minutes. If you find any of the points unsatisfactory, call the responsible person and ask him to repeat the investigation till he can guarantee that a repetition of the same or a similar accident can be definitely avoided.

For your next safety visit to the section, make a note of the accident and ask the supervisor to take you to the injured person, talk with him about his accident and how he feels now, and whether he was satisfied with the treatment.

<table>
<tr><td colspan="2">Accident Investigation Report</td><td>Plant: Product Flow</td><td>Section: Tank farm</td></tr>
<tr><td colspan="4">Accident No. 321PF</td></tr>
<tr><td>Date: 13 March 2006</td><td colspan="2">Time: 14:50</td><td>Weekday: Monday</td></tr>
<tr><td>Job of injured person</td><td colspan="3">Maintenance mechanic</td></tr>
<tr><td>Type of injury</td><td colspan="3">Cut in left leg</td></tr>
<tr><td>Treated where</td><td colspan="3">Town hospital</td></tr>
<tr><td>Expected days of absence</td><td colspan="2">maybe a week</td><td>Recordable? Yes</td></tr>
<tr><td colspan="3">Preliminary information distributed on: 15 March 2006</td><td>at: 11:00</td></tr>
<tr><td colspan="4">What happened

The mechanic did not use the hook to remove the lid over an oil separator ditch, and used too much force to remove it with his hands, as it was jammed. The lid suddenly gave way, and the mechanic, who had put his foot too close to the opening, hurt himself. Other workers have also complained that the lid is too heavy, and this was also known to the mechanic. He had been often seen opening the lids without using the hook, and had also been warned to stop such practice, as it was against the procedure.</td></tr>
<tr><td colspan="4">Cause of accident

In spite of having been told a few times, the mechanic carried out an unsafe act which resulted in this injury. He does not seem to want to follow the procedure prescribed for this job. Knowing that the lid was heavy, it was careless on his part to have attempted to remove the lid in this manner.</td></tr>
</table>

Countermeasures	Person	Deadline
1. The mechanic to be told once more not to repeat the unsafe act, or to expect a written warning	B. Wooster	after his return
2. Description of the accident with similar warning to be distributed to all operators and mechanics	B. Little	3rd week March
3. A notice describing the hazards of carelessness, with special reference to this incident, to be put on all notice-boards	H. Sicherle	asap

Investigation Team (person, position)

1. B. Wooster, Supervisor tank farm
2. H. Sicherle, Safety Manager for the site
3. B. Little, plant safety person

Start of investigation	End of investigation	This report date
20. 03. 06	20. 03. 06	29. 03. 06

Distribution: Works Manager, Plant Manager, Plant Safety Person, Central Safety Department, Notice Board

Appendix 5:
About *audit* and audits

The word "audit", as mentioned, is derived from the Latin verb *audire* = to hear. *Audit* is third person singular of *audire,* and means "he, she, it hears". There are several words that are derived from *audire,* for example "audible" = something that can be heard, "audience" = hearers, "audition" = hearing for a trial, "auditorium" = a hall to hear something, and of course, "audio" together with "video", the latter coming similarly from *videre* = to see.

In the 17th century, grades of oral examinations in universities like Oxford were communicated to the students orally. The student "heard" his grades, i.e. "he hears" = *audit.* Since then, the word audit is associated with examinations. It was initially taken up by the official examiners of accounts, the auditors. Since then, "official" examinations of all kinds, be it by the government authorities, or corporate authorities are called audits. As I do not want managers to go around *examining* their people, I have replaced the word "audit" by the word "visit".

Auditing, as it is understood by auditing and certifying companies today, is an exercise whereby one has a set of statements or activities, and the auditor checks whether they are fulfilled. Audits are also carried out by certifying companies from whom companies seek certifications like ISO 9000. Big consulting companies have long questionnaires, where the auditor only ticks in the appropriate column whether the answer given to him matches the statements' demands. The real work is in constructing the questionnaire, which is done by experienced specialists. The filling-out of the forms can be done by anybody. It does not require any special knowledge, because the answers have to be only "Yes" or "No". This is the reason one is flabbergasted when a "senior" consultant from such consulting companies, about 25 years old, comes and starts evaluating whether your job,

Managing Safety. Kishor Bhagwati

ISBN 3-527-31583-7

which is built up on years of knowledge and experience, is necessary for the company or not.

August Horch, born 1868 near Koblenz in Germany, was appointed in 1896 as works manager by Carl Benz, who had an automobile factory in Mannheim near Heidelberg. Carl Benz & Cie. later merged with Daimler Motorengesellschaft in Stuttgart to create Daimler-Benz, which make Mercedes cars. Horch founded his own factory to make cars, called Horch & Co in Zwickau in Saxony in Germany. Having had a disagreement with his partners, he separated from the company and founded a new company. Not being able to use the Horch name for his new company, because the former company wanted to keep the name, he decided to open it under another name.

"Horch" in German means "hark" or "listen" or "hear". He named the new company also "Horch", but in Latin. Can you guess what name he gave to his company? Yes, Audi. Later the former Horch and Audi companies merged with two other automobile companies in Saxony to form the Auto-Union. The logo they selected for the new company formed out of the four showed four intertwined rings. This logo is seen on Audi cars today.

Coming back to audits, at the time the colleges in England pronounced their examination results, a special strong ale was brewed by the breweries, called "Audit Ale" for the students, either to celebrate their success or to comfort those who had failed. Audit Ale was also

Figure 19 Audit Ale

given traditionally by English lords to their tenant farmers when the farmers' land rents were due.

Tim Alborn has made a nice limerick on audit ale:

The English might merit a plaudit
For inventing the ale they call audit.
This beer was ingested
When students were tested;
Both winners and losers would laud it.

Appendix 6: Safety visit Control Sheet

The following sheet enables the manager to check whether his subordinates are carrying out their safety visits as per the schedule that has been put up by the safety department in collaboration with the managers and you. This sheet, drawn up for the past 12 months, is to be presented every quarter by the safety professional at the management board meeting, and those who have not fulfilled their quota have to answer for their irregularities in carrying out safety visits.

The following table is for the managerial level of the simple organisation given in Chapter 13.

	J	F	M	A	M	J	J	A	S	O	N	D
Works Manager												
Manager Plant 1												
Manager Plant 2												
Manager Finance												
Manager HR												

One can just put a cross in the box to show that the safety visit was carried out, as follows:

	J	F	M	A	M	J	J	A	S	O	N	D
Works Manager	x	x	x	x	x	x		x	x	x	x	x
Manager Plant 1	x	x				x		x		x	x	
Manager Plant 2	x	x	x	x		x	x	x	x	x	x	x
Manager Finance			x						x			
Manager HR		x						x			xx	

Managing Safety. Kishor Bhagwati

ISBN 3-527-31583-7

The table shows that Manager Plant 1 was not carrying out his visits as required, whereas Manager Plant 2 had almost fulfilled his quota. The HR Manager also fulfilled his quota by making up for the safety visit he missed out in May with two visits in December. This may be allowed as an exception, but should not become the rule.

A further step is putting the act to condition ratio in the boxes, with ratios below 1 being marked in red (or in italics as here):

	J	F	M	A	M	J	J	A	S	O	N	D
Works Manager	1.3	1.2	1.6	1.5	1.5	*0.9*		1.8	1.2	1.1	1.4	1.5
Manager Plant 1	1.0	*0.7*				*0.3*		*0.8*		1.1	*0.3*	
Manager Plant 2	1.4	1.7	2.0	*0.9*		1.6	1.8	1.4	*0.7*	1.3	1.1	1.5
Manager Finance			1.1						*0.8*			
Manager HR		1.4						1.6			1.2	

It seems that Manager Plant 1 has not yet understood the purpose of safety visits and needs coaching in it. If the coaching does not improve his ratio, the works manager should have a serious talk with him. *All managers should be made aware that their performance here will have a considerable influence on their annual appraisal.*

Index

Managing Safety. Kishor Bhagwati

ISBN 3-527-31583-7